Laser: Science and Engineering

Laser: Science and Engineering

Edited by
Kayden Robertson

⊟ Larsen & Keller
www.larsen-keller.com

Laser: Science and Engineering
Edited by Kayden Robertson
ISBN: 978-1-63549-164-7 (Hardback)

© 2017 Larsen & Keller

☰ Larsen & Keller

Published by Larsen and Keller Education,
5 Penn Plaza,
19th Floor,
New York, NY 10001, USA

Cataloging-in-Publication Data

Laser : science and engineering / edited by Kayden Robertson.
   p. cm.
Includes bibliographical references and index.
ISBN 978-1-63549-164-7
1. Lasers. 2. Lasers in engineering. 3. Lasers--Industrial applications. I. Robertson, Kayden.
TA1675 .L37 2017
621.366--dc23

This book contains information obtained from authentic and highly regarded sources. All chapters are published with permission under the Creative Commons Attribution Share Alike License or equivalent. A wide variety of references are listed. Permissions and sources are indicated; for detailed attributions, please refer to the permissions page. Reasonable efforts have been made to publish reliable data and information, but the authors, editors and publisher cannot assume any responsibility for the vailidity of all materials or the consequences of their use.

Trademark Notice: All trademarks used herein are the property of their respective owners. The use of any trademark in this text does not vest in the author or publisher any trademark ownership rights in such trademarks, nor does the use of such trademarks imply any affiliation with or endorsement of this book by such owners.

The publisher's policy is to use permanent paper from mills that operate a sustainable forestry policy. Furthermore, the publisher ensures that the text paper and cover boards used have met acceptable environmental accreditation standards.

Printed and bound in the United States of America.

For more information regarding Larsen and Keller Education and its products, please visit the publisher's website www.larsen-keller.com

Table of Contents

Permissions

Index

Preface

This book is compiled in such a manner, that it will provide in-depth knowledge about the theory and practice of laser and laser technology. It will give deep insights about the applications of laser and its technology in fields like science, medical and engineering. Laser is an acronym for light amplification by stimulated emission of radiation. It is used in barcode scanners, optical disk layers, laser surgery, laser lightning displays, etc. This textbook presents the complex subject of lasers in the most comprehensible and easy to understand language. Most of the topics introduced in it cover new techniques and the applications of laser technology. Different approaches, evaluations and methodologies have been included in it. The text seeks to touch upon the various applications and technological advances related to lasers. It will serve as a valuable source of reference for those interested in this subject.

A foreword of all Chapters of the book is provided below:

Chapter 1 - Lasers are amplified light and are characterized by their wavelength in a vacuum. It is one of the most significant technologies that have been manufactured in the last century. This chapter provides an introduction to lasers, their fundamental characteristics, design and related terminologies as well as the basic theories related to its manufacture.; **Chapter 2 -** Lasers can be classified based on the gain medium used to amplify light. They can be classified as gas laser, chemical laser, dye laser, solid-state laser, gas dynamic laser, free-electron laser, Raman laser, nuclear pumped laser, excimer laser, fiber laser etc. Gas laser can further be divided into nitrogen laser, helium-neon laser and ion laser. This chapter describes each of these lasers in detail explaining their discerning features and working mechanisms.; **Chapter 3 -** There are several techniques by which lasers operate to produce light. This chapter describes in detail the techniques of Q-switching, mode-locking and gain-switching. This content also includes manufacturing and marking techniques like laser bonding, laser cutting, direct metal laser sintering and laser diffraction analysis. These techniques help in understanding the way in which lasers operate and also the uses of these techniques.; **Chapter 4 -** There are numerous gadgets that use laser technology in their workings. This chapter examines devices like optical disc drive, laser printers, 3D scanner, barcode readers, magneto-optical trap and laser pointer and techniques like laser-based angle-resolved photoemission spectroscopy, fiber-optic communication etc. The section details the applications and design of such devices and techniques. The topics discussed in the chapter are of great importance to broaden the existing knowledge on laser.; **Chapter 5 -** The usage of lasers in industrial manufacturing is multifold and this chapter explores laser cooling, laser ablation, optical amplifier, optical tweezers, laser beam welding, laser beam machining and chirped pulse amplification. The various industries that use lasers include the military, aerospace and aeronautics, the security industry, the medical industry and many others. The chapter describes the type of lasers used and the application of these processes.

I would like to thank the entire editorial team who made sincere efforts for this book and my family who supported me in my efforts of working on this book. I take this opportunity to thank all those who have been a guiding force throughout my life.

Editor

Introduction to Laser

Lasers are amplified light and are characterized by their wavelength in a vacuum. It is one of the most significant technologies that have been manufactured in the last century. This chapter provides an introduction to lasers, their fundamental characteristics, design and related terminologies as well as the basic theories related to its manufacture.

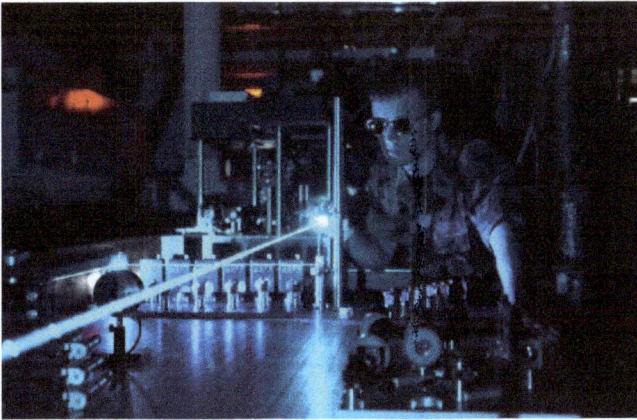

United States Air Force laser experiment

A laser is a device that emits light through a process of optical amplification based on the stimulated emission of electromagnetic radiation. The term "laser" originated as an acronym for "light amplification by stimulated emission of radiation". The first laser was built in 1960 by Theodore H. Maiman at Hughes Research Laboratories, based on theoretical work by Charles Hard Townes and Arthur Leonard Schawlow. A laser differs from other sources of light in that it emits light *coherently*. Spatial coherence allows a laser to be focused to a tight spot, enabling applications such as laser cutting and lithography. Spatial coherence also allows a laser beam to stay narrow over great distances (collimation), enabling applications such as laser pointers. Lasers can also have high temporal coherence, which allows them to emit light with a very narrow spectrum, i.e., they can emit a single color of light. Temporal coherence can be used to produce pulses of light as short as a femtosecond.

Among their many applications, lasers are used in optical disk drives, laser printers, and barcode scanners; DNA sequencing instruments, fiber-optic and free-space optical communication; laser surgery and skin treatments; cutting and welding materials; military and law enforcement devices for marking targets and measuring range and speed; and laser lighting displays in entertainment.

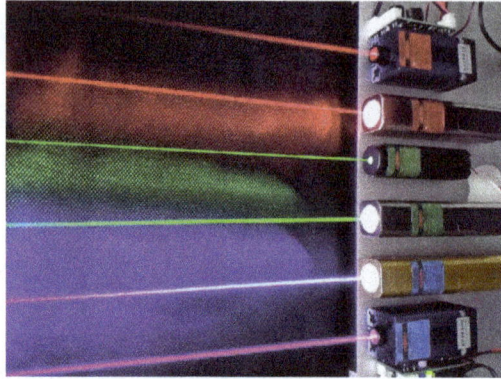

Red (660 & 635 nm), green (532 & 520 nm) and blue-violet (445 & 405 nm) lasers

Fundamentals

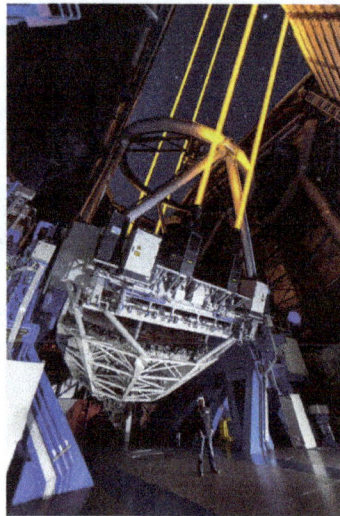

Modern telescopes use laser technologies to compensate for the blurring effect of the Earth's atmosphere.

Lasers are distinguished from other light sources by their coherence. Spatial coherence is typically expressed through the output being a narrow beam, which is diffraction-limited. Laser beams can be focused to very tiny spots, achieving a very high irradiance, or they can have very low divergence in order to concentrate their power at a great distance.

Temporal (or longitudinal) coherence implies a polarized wave at a single frequency whose phase is correlated over a relatively great distance (the coherence length) along the beam. A beam produced by a thermal or other incoherent light source has an instantaneous amplitude and phase that vary randomly with respect to time and position, thus having a short coherence length.

Lasers are characterized according to their wavelength in a vacuum. Most "single wavelength" lasers actually produce radiation in several *modes* having slightly differing frequencies (wavelengths), often not in a single polarization. Although temporal coherence implies monochromaticity, there are lasers that emit a broad spectrum of light or emit different wavelengths of light simultaneously. There are some lasers that are not single spatial mode and consequently have light beams that diverge more than is required by the diffraction limit. However, all such devices are classified as "lasers" based on their method of producing light, i.e., stimulated emission. Lasers are employed in applications where light of the required spatial or temporal coherence could not be produced using simpler technologies.

Terminology

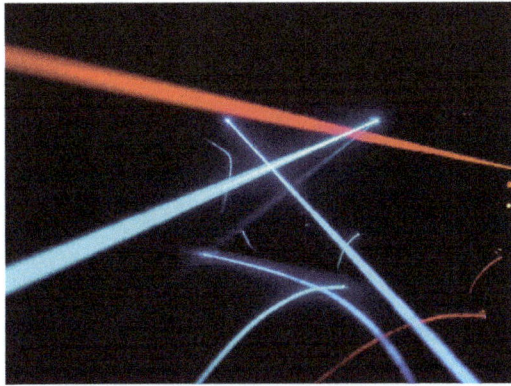

Laser beams in fog, reflected on a car windshield

The word *laser* started as an acronym for "light amplification by stimulated emission of radiation". In modern usage, the term "light" includes electromagnetic radiation of any frequency, not only visible light, hence the terms *infrared laser*, *ultraviolet laser*, *X-ray laser*, *gamma-ray laser*, and so on. Because the microwave predecessor of the laser, the maser, was developed first, devices of this sort operating at microwave and radio frequencies are referred to as "masers" rather than "microwave lasers" or "radio lasers". In the early technical literature, especially at Bell Telephone Laboratories, the laser was called an optical maser; this term is now obsolete.

A laser that produces light by itself is technically an optical oscillator rather than an optical amplifier as suggested by the acronym. It has been humorously noted that the acronym LOSER, for "light oscillation by stimulated emission of radiation", would have been more correct. With the widespread use of the original acronym as a common noun, optical amplifiers have come to be referred to as "laser amplifiers", notwithstanding the apparent redundancy in that designation.

The back-formed verb *to lase* is frequently used in the field, meaning "to produce laser light," especially in reference to the gain medium of a laser; when a laser is operating it is said to be "lasing." Further use of the words *laser* and *maser* in an extended sense, not referring to laser technology or devices, can be seen in usages such as *astrophysical maser* and *atom laser*.

Design

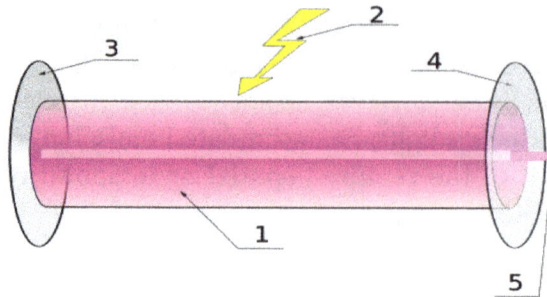

Components of a typical laser:
1. Gain medium, 2. Laser pumping energy, 3. High reflector, 4. Output coupler, 5. Laser beam

Animation explaining stimulated emission and the laser principle

A laser consists of a gain medium, a mechanism to energize it, and something to provide optical feedback. The gain medium is a material with properties that allow it to amplify light by way of stimulated emission. Light of a specific wavelength that passes through the gain medium is amplified (increases in power).

For the gain medium to amplify light, it needs to be supplied with energy in a process called pumping. The energy is typically supplied as an electric current or as light at a different wavelength. Pump light may be provided by a flash lamp or by another laser.

The most common type of laser uses feedback from an optical cavity—a pair of mirrors on either end of the gain medium. Light bounces back and forth between the mirrors, passing through the gain medium and being amplified each time. Typically one of the two mirrors, the output coupler, is partially transparent. Some of the light escapes through this mirror. Depending on the design of the cavity (whether the mirrors are flat or curved), the light coming out of the laser may spread out or form a narrow beam. In analogy to electronic oscillators, this device is sometimes called a *laser oscillator*.

Most practical lasers contain additional elements that affect properties of the emitted light, such as the polarization, wavelength, and shape of the beam.

Laser Physics

Electrons and how they interact with electromagnetic fields are important in our understanding of chemistry and physics.

Stimulated Emission

In the classical view, the energy of an electron orbiting an atomic nucleus is larger for orbits further from the nucleus of an atom. However, quantum mechanical effects force electrons to take on discrete positions in orbitals. Thus, electrons are found in specific energy levels of an atom, two of which are shown below:

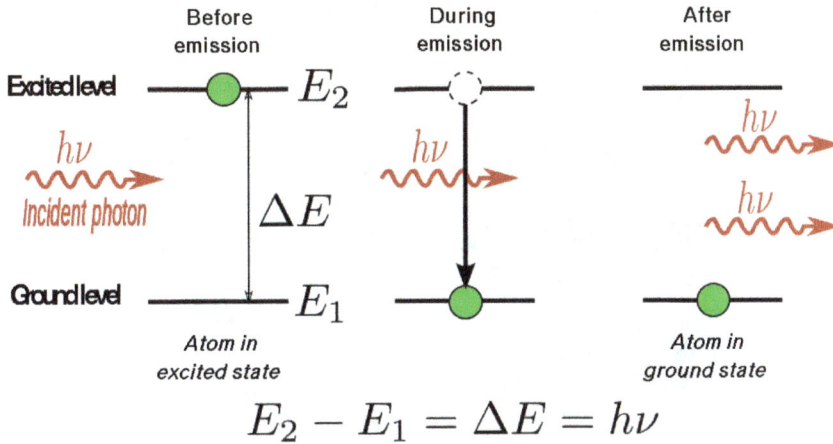

$$E_2 - E_1 = \Delta E = h\nu$$

When an electron absorbs energy either from light (photons) or heat (phonons), it receives that incident quantum of energy. But transitions are only allowed in between discrete energy levels such as the two shown above. This leads to emission lines and absorption lines.

When an electron is excited from a lower to a higher energy level, it will not stay that way forever. An electron in an excited state may decay to a lower energy state which is not occupied, according to a particular time constant characterizing that transition. When such an electron decays without external influence, emitting a photon, that is called "spontaneous emission". The phase associated with the photon that is emitted is random. A material with many atoms in such an excited state may thus result in radiation which is very spectrally limited (centered around one wavelength of light), but the individual photons would have no common phase relationship and would emanate in random directions. This is the mechanism of fluorescence and thermal emission.

An external electromagnetic field at a frequency associated with a transition can affect the quantum mechanical state of the atom. As the electron in the atom makes a transition between two stationary states (neither of which shows a dipole field), it enters a transition state which does have a dipole field, and which acts like a small electric dipole, and this dipole oscillates at a characteristic frequency. In response to the external electric field at this frequency, the probability of the atom entering this transition state is greatly increased. Thus, the rate of transitions between two stationary states is enhanced beyond that due to spontaneous emission. Such a transition to the higher state is called absorption, and it destroys an incident photon (the photon's energy goes into powering the increased energy of the higher state). A transition from the higher to a lower energy state, however, produces an additional photon; this is the process of stimulated emission.

Gain Medium and Cavity

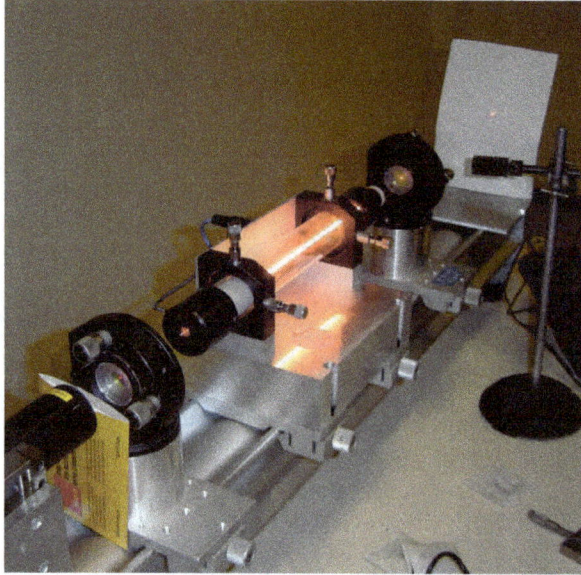

A helium–neon laser demonstration at the Kastler-Brossel Laboratory at Univ. Paris 6. The pink-orange glow running through the center of the tube is from the electric discharge which produces incoherent light, just as in a neon tube. This glowing plasma is excited and then acts as the gain medium through which the internal beam passes, as it is reflected between the two mirrors. Laser radiation output through the front mirror can be seen to produce a tiny (about 1 mm in diameter) intense spot on the screen, to the right. Although it is a deep and pure red color, spots of laser light are so intense that cameras are typically overexposed and distort their color.

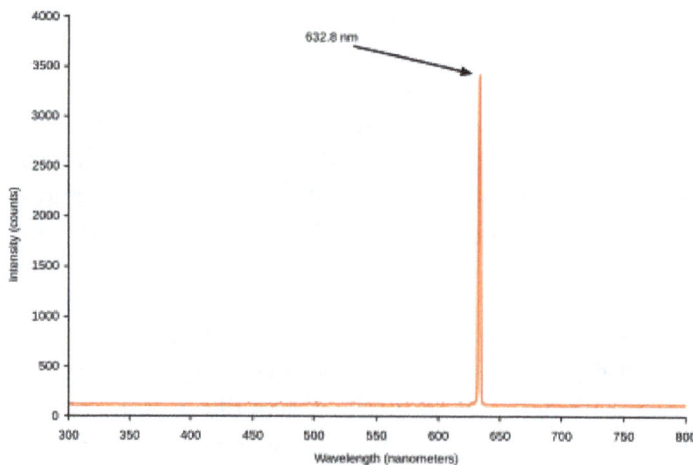

Spectrum of a helium neon laser illustrating its very high spectral purity (limited by the measuring apparatus). The 0.002 nm bandwidth of the lasing medium is well over 10,000 times narrower than the spectral width of a light-emitting diode (whose spectrum is shown here for comparison), with the bandwidth of a single longitudinal mode being much narrower still.

The gain medium is excited by an external source of energy into an excited state. In most lasers this medium consists of a population of atoms which have been excited into such a state by means of an outside light source, or an electrical field which supplies energy for atoms to absorb and be transformed into their excited states.

The gain medium of a laser is normally a material of controlled purity, size, concentration, and shape, which amplifies the beam by the process of stimulated emission described above. This material can be of any state: gas, liquid, solid, or plasma. The gain medium absorbs pump energy, which raises some electrons into higher-energy ("excited") quantum states. Particles can interact with light by either absorbing or emitting photons. Emission can be spontaneous or stimulated. In the latter case, the photon is emitted in the same direction as the light that is passing by. When the number of particles in one excited state exceeds the number of particles in some lower-energy state, population inversion is achieved and the amount of stimulated emission due to light that passes through is larger than the amount of absorption. Hence, the light is amplified. By itself, this makes an optical amplifier. When an optical amplifier is placed inside a resonant optical cavity, one obtains a laser oscillator.

In a few situations it is possible to obtain lasing with only a single pass of EM radiation through the gain medium, and this produces a laser beam without any need for a resonant or reflective cavity (see for example nitrogen laser). Thus, reflection in a resonant cavity is usually required for a laser, but is not absolutely necessary.

The optical resonator is sometimes referred to as an "optical cavity", but this is a misnomer: lasers use open resonators as opposed to the literal cavity that would be employed at microwave frequencies in a maser. The resonator typically consists of two mirrors between which a coherent beam of light travels in both directions, reflecting back on itself so that an average photon will pass through the gain medium repeatedly before it is emitted from the output aperture or lost to diffraction or absorption. If the gain (amplification) in the medium is larger than the resonator losses, then the power of the recirculating light can rise exponentially. But each stimulated emission event returns an atom from its excited state to the ground state, reducing the gain of the medium. With increasing beam power the net gain (gain minus loss) reduces to unity and the gain medium is said to be saturated. In a continuous wave (CW) laser, the balance of pump power against gain saturation and cavity losses produces an equilibrium value of the laser power inside the cavity; this equilibrium determines the operating point of the laser. If the applied pump power is too small, the gain will never be sufficient to overcome the resonator losses, and laser light will not be produced. The minimum pump power needed to begin laser action is called the *lasing threshold*. The gain medium will amplify any photons passing through it, regardless of direction; but only the photons in a spatial mode supported by the resonator will pass more than once through the medium and receive substantial amplification.

The Light Emitted

The light generated by stimulated emission is very similar to the input signal in terms of wavelength, phase, and polarization. This gives laser light its characteristic coherence, and allows it to maintain the uniform polarization and often monochromaticity established by the optical cavity design.

The beam in the cavity and the output beam of the laser, when travelling in free space (or a homogeneous medium) rather than waveguides (as in an optical fiber laser), can be approximated as a Gaussian beam in most lasers; such beams exhibit the minimum divergence for a given diameter. However some high power lasers may be multimode, with the transverse modes often approximated using Hermite–Gaussian or Laguerre-Gaussian functions. It has been shown that unstable laser resonators (not used in most lasers) produce fractal shaped beams. Near the beam "waist" (or focal region) it is highly *collimated*: the wavefronts are planar, normal to the direction of propagation, with no beam divergence at that point. However, due to diffraction, that can only remain true well within the Rayleigh range. The beam of a single transverse mode (gaussian beam) laser eventually diverges at an angle which varies inversely with the beam diameter, as required by diffraction theory. Thus, the "pencil beam" directly generated by a common helium–neon laser would spread out to a size of perhaps 500 kilometers when shone on the Moon (from the distance of the earth). On the other hand, the light from a semiconductor laser typically exits the tiny crystal with a large divergence: up to 50°. However even such a divergent beam can be transformed into a similarly collimated beam by means of a lens system, as is always included, for instance, in a laser pointer whose light originates from a laser diode. That is possible due to the light being of a single spatial mode. This unique property of laser light, spatial coherence, cannot be replicated using standard light sources (except by discarding most of the light) as can be appreciated by comparing the beam from a flashlight (torch) or spotlight to that of almost any laser.

Quantum Vs. Classical Emission Processes

The mechanism of producing radiation in a laser relies on stimulated emission, where energy is extracted from a transition in an atom or molecule. This is a quantum phenomenon discovered by Einstein who derived the relationship between the A coefficient describing spontaneous emission and the B coefficient which applies to absorption and stimulated emission. However, in the case of the free electron laser, atomic energy levels are not involved; it appears that the operation of this rather exotic device can be explained without reference to quantum mechanics.

Continuous and Pulsed Modes of Operation

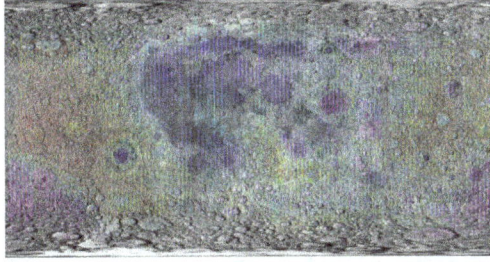
Lidar measurements of lunar topography made by Clementine mission.

Laserlink point to point optical wireless network

A laser can be classified as operating in either continuous or pulsed mode, depending on whether the power output is essentially continuous over time or whether its output takes the form of pulses of light on one or another time scale. Of course even a laser whose output is normally continuous can be intentionally turned on and off at some rate in order to create pulses of light. When the modulation rate is on time scales much slower than the cavity lifetime and the time period over which energy can be stored in the lasing medium or pumping mechanism, then it is still classified as a "modulated" or "pulsed" continuous wave laser. Most laser diodes used in communication systems fall in that category.

Mercury Laser Altimeter (MLA) of the MESSENGER spacecraft

Continuous Wave Operation

Some applications of lasers depend on a beam whose output power is constant over time. Such a laser is known as *continuous wave* (*CW*). Many types of lasers can be made to operate in continuous wave mode to satisfy such an application. Many of these lasers actually lase in several longitudinal modes at the same time, and beats between the slightly different optical frequencies of those oscillations will in fact produce amplitude variations on time scales shorter than the round-trip time (the reciprocal of the frequency spacing between modes), typically a few nanoseconds or less. In most cases these lasers are still termed "continuous wave" as their output power is steady when averaged over any longer time periods, with the very high frequency power variations having little or no impact in the intended application. (However the term is not applied to mode-locked lasers, where the *intention* is to create very short pulses at the rate of the round-trip time).

For continuous wave operation it is required for the population inversion of the gain medium to be continually replenished by a steady pump source. In some lasing media this is impossible. In some other lasers it would require pumping the laser at a very high continuous power level which would be impractical or destroy the laser by producing excessive heat. Such lasers cannot be run in CW mode.

Pulsed Operation

Pulsed operation of lasers refers to any laser not classified as continuous wave, so that the optical power appears in pulses of some duration at some repetition rate. This encompasses a wide range of technologies addressing a number of different motivations. Some lasers are pulsed simply because they cannot be run in continuous mode.

In other cases the application requires the production of pulses having as large an energy as possible. Since the pulse energy is equal to the average power divided by the repetition rate, this goal can sometimes be satisfied by lowering the rate of pulses so that more energy can be built up in between pulses. In laser ablation for example, a small volume of material at the surface of a work piece can be evaporated if it is heated in a very short time, whereas supplying the energy gradually would allow for the heat to be absorbed into the bulk of the piece, never attaining a sufficiently high temperature at a particular point.

Other applications rely on the peak pulse power (rather than the energy in the pulse), especially in order to obtain nonlinear optical effects. For a given pulse energy, this requires creating pulses of the shortest possible duration utilizing techniques such as Q-switching.

The optical bandwidth of a pulse cannot be narrower than the reciprocal of the pulse width. In the case of extremely short pulses, that implies lasing over a considerable bandwidth, quite contrary to the very narrow bandwidths typical of CW lasers. The lasing medium in some *dye lasers* and *vibronic solid-state lasers* produces optical gain over a wide bandwidth, making a laser possible which can thus generate pulses of light as short as a few femtoseconds (10^{-15} s).

Q-Switching

In a Q-switched laser, the population inversion is allowed to build up by introducing loss inside the resonator which exceeds the gain of the medium; this can also be described as a reduction of the quality factor or 'Q' of the cavity. Then, after the pump energy stored in the laser medium has approached the maximum possible level, the introduced loss mechanism (often an electro- or acousto-optical element) is rapidly removed (or that occurs by itself in a passive device), allowing lasing to begin which rapidly obtains the stored energy in the gain medium. This results in a short pulse incorporating that energy, and thus a high peak power.

Mode-Locking

A mode-locked laser is capable of emitting extremely short pulses on the order of tens of picoseconds down to less than 10 femtoseconds. These pulses will repeat at the round trip time, that is, the time that it takes light to complete one round trip between the mirrors comprising the resonator. Due to the Fourier limit (also known as energy-time uncertainty), a pulse of such short temporal length has a spectrum spread over a considerable bandwidth. Thus such a gain medium must have a gain bandwidth sufficiently broad to amplify those frequencies. An example of a suitable material is titanium-doped, artificially grown sapphire (Ti:sapphire) which has a very wide gain bandwidth and can thus produce pulses of only a few femtoseconds duration.

Such mode-locked lasers are a most versatile tool for researching processes occurring on extremely short time scales (known as femtosecond physics, femtosecond chemistry and ultrafast science), for maximizing the effect of nonlinearity in optical materials (e.g. in second-harmonic generation, parametric down-conversion, optical parametric oscillators and the like) due to the large peak power, and in ablation applications. Again, because of the extremely short pulse duration, such a laser will produce pulses which achieve an extremely high peak power.

Pulsed Pumping

Another method of achieving pulsed laser operation is to pump the laser material with a source that is itself pulsed, either through electronic charging in the case of flash lamps, or another laser which is already pulsed. Pulsed pumping was historically used with dye lasers where the inverted population lifetime of a dye molecule was so short that a high energy, fast pump was needed. The way to overcome this problem was to charge up large capacitors which are then switched to discharge through flashlamps, producing an intense flash. Pulsed pumping is also required for three-level lasers in which the lower energy level rapidly becomes highly populated preventing further lasing until those atoms relax to the ground state. These lasers, such as the excimer laser and the copper vapor laser, can never be operated in CW mode.

History

Foundations

In 1917, Albert Einstein established the theoretical foundations for the laser and the maser in the paper *Zur Quantentheorie der Strahlung* (On the Quantum Theory of Radiation) via a re-derivation of Max Planck's law of radiation, conceptually based upon probability coefficients (Einstein coefficients) for the absorption, spontaneous emission, and stimulated emission of electromagnetic radiation. In 1928, Rudolf W. Ladenburg confirmed the existence of the phenomena of stimulated emission and negative absorption. In 1939, Valentin A. Fabrikant predicted the use of stimulated emission to amplify "short" waves. In 1947, Willis E. Lamb and R. C. Retherford found apparent stimulated emission in hydrogen spectra and effected the first demonstration of stimulated emission. In 1950, Alfred Kastler (Nobel Prize for Physics 1966) proposed the method of optical pumping, experimentally confirmed, two years later, by Brossel, Kastler, and Winter.

Maser

In 1951, Joseph Weber submitted a paper on using stimulated emissions to make a microwave amplifier to the June 1952 Institute of Radio Engineers Vacuum Tube Research Conference at Ottawa. After this presentation, RCA asked Weber to give a seminar on this idea, and Charles Hard Townes asked him for a copy of the paper.

Aleksandr Prokhorov

In 1953, Charles Hard Townes and graduate students James P. Gordon and Herbert J. Zeiger produced the first microwave amplifier, a device operating on similar principles to the laser, but amplifying microwave radiation rather than infrared or visible radiation. Townes's maser was incapable of continuous output. Meanwhile, in the Soviet Union, Nikolay Basov and Aleksandr Prokhorov were independently working on the quantum oscillator and solved the problem of continuous-output systems by using more than two energy levels. These gain media could release stimulated emissions between an excited state and a lower excited state, not the ground state, facilitating the maintenance of a population inversion. In 1955, Prokhorov and Basov suggested optical pumping of a multi-level system as a method for obtaining the population inversion, later a main method of laser pumping.

Townes reports that several eminent physicists—among them Niels Bohr, John von Neumann, and Llewellyn Thomas—argued the maser violated Heisenberg's uncertainty principle and hence could not work. Others such as Isidor Rabi and Polykarp Kusch expected that it would be impractical and not worth the effort. In 1964 Charles H. Townes, Nikolay Basov, and Aleksandr Prokhorov shared the Nobel Prize in Physics, "for fundamental work in the field of quantum electronics, which has led to the construction of oscillators and amplifiers based on the maser–laser principle".

Laser

In 1957, Charles Hard Townes and Arthur Leonard Schawlow, then at Bell Labs, began a serious study of the infrared laser. As ideas developed, they abandoned infrared radiation to instead concentrate upon visible light. The concept originally was called an "optical maser". In 1958, Bell Labs filed a patent application for their proposed optical maser; and Schawlow and Townes submitted a manuscript of their theoretical calculations to the *Physical Review*, published that year in Volume 112, Issue No. 6.

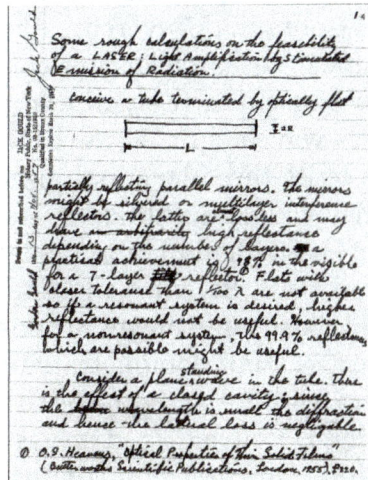

LASER notebook: First page of the notebook wherein Gordon Gould coined the LASER acronym, and described the elements for constructing the device.

Simultaneously, at Columbia University, graduate student Gordon Gould was work-ing on a doctoral thesis about the energy levels of excited thallium. When Gould and Townes met, they spoke of radiation emission, as a general subject; afterwards, in November 1957, Gould noted his ideas for a "laser", including using an open reso-nator (later an essential laser-device component). Moreover, in 1958, Prokhorov in-dependently proposed using an open resonator, the first published appearance (the USSR) of this idea. Elsewhere, in the U.S., Schawlow and Townes had agreed to an open-resonator laser design – apparently unaware of Prokhorov's publications and Gould's unpublished laser work.

At a conference in 1959, Gordon Gould published the term LASER in the paper *The LASER, Light Amplification by Stimulated Emission of Radiation*. Gould's linguistic intention was using the "-aser" word particle as a suffix – to accurately denote the spec-trum of the light emitted by the LASER device; thus x-rays: *xaser*, ultraviolet: *uvaser*, et cetera; none established itself as a discrete term, although "raser" was briefly popular for denoting radio-frequency-emitting devices.

Gould's notes included possible applications for a laser, such as spectrometry, interfer-ometry, radar, and nuclear fusion. He continued developing the idea, and filed a patent application in April 1959. The U.S. Patent Office denied his application, and awarded a patent to Bell Labs, in 1960. That provoked a twenty-eight-year lawsuit, featuring scientific prestige and money as the stakes. Gould won his first minor patent in 1977, yet it was not until 1987 that he won the first significant patent lawsuit victory, when a Federal judge ordered the U.S. Patent Office to issue patents to Gould for the optically pumped and the gas discharge laser devices. The question of just how to assign credit for inventing the laser remains unresolved by historians.

On May 16, 1960, Theodore H. Maiman operated the first functioning laser at Hughes Research Laboratories, Malibu, California, ahead of several research teams, including those of Townes, at Columbia University, Arthur Schawlow, at Bell Labs, and Gould, at the TRG (Technical Research Group) company. Maiman's functional laser used a solid-state flashlamp-pumped synthetic ruby crystal to produce red laser light, at 694 nanometers wavelength; however, the device only was capable of pulsed operation, because of its three-level pumping design scheme. Later that year, the Iranian physicist Ali Javan, and William R. Bennett, and Donald Herriott, constructed the first gas laser, using helium and neon that was capable of contin-uous operation in the infrared (U.S. Patent 3,149,290); later, Javan received the Albert Einstein Award in 1993. Basov and Javan proposed the semiconductor laser diode concept. In 1962, Robert N. Hall demonstrated the first *laser diode* device, made of gallium arsenide and emitted at 850 nm the near-infrared band of the spectrum. Later that year, Nick Holonyak, Jr. demonstrated the first semiconduc-tor laser with a visible emission. This first semiconductor laser could only be used in pulsed-beam operation, and when cooled to liquid nitrogen temperatures (77 K). In 1970, Zhores Alferov, in the USSR, and Izuo Hayashi and Morton Panish of Bell

Telephone Laboratories also independently developed room-temperature, continual-operation diode lasers, using the heterojunction structure.

Recent Innovations

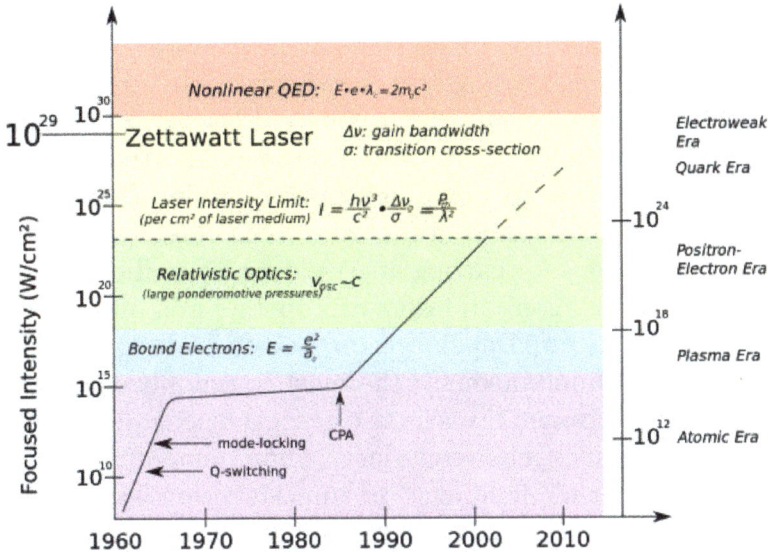

Graph showing the history of maximum laser pulse intensity throughout the past 40 years.

Since the early period of laser history, laser research has produced a variety of improved and specialized laser types, optimized for different performance goals, including:

- new wavelength bands
- maximum average output power
- maximum peak pulse energy
- maximum peak pulse power
- minimum output pulse duration
- maximum power efficiency
- minimum cost

and this research continues to this day.

Lasing without maintaining the medium excited into a population inversion was discovered in 1992 in sodium gas and again in 1995 in rubidium gas by various international teams. This was accomplished by using an external maser to induce "optical transparency" in the medium by introducing and destructively interfering the ground electron transitions between two paths, so that the likelihood for the ground electrons to absorb any energy has been cancelled.

Types and Operating Principles

Gas Lasers

Following the invention of the HeNe gas laser, many other gas discharges have been found to amplify light coherently. Gas lasers using many different gases have been built and used for many purposes. The helium–neon laser (HeNe) is able to operate at a number of different wavelengths, however the vast majority are engineered to lase at 633 nm; these relatively low cost but highly coherent lasers are extremely common in optical research and educational laboratories. Commercial carbon dioxide (CO_2) lasers can emit many hundreds of watts in a single spatial mode which can be concentrated into a tiny spot. This emission is in the thermal infrared at 10.6 μm; such lasers are regularly used in industry for cutting and welding. The efficiency of a CO_2 laser is unusually high: over 30%. Argon-ion lasers can operate at a number of lasing transitions between 351 and 528.7 nm. Depending on the optical design one or more of these transitions can be lasing simultaneously; the most commonly used lines are 458 nm, 488 nm and 514.5 nm. A nitrogen transverse electrical discharge in gas at atmospheric pressure (TEA) laser is an inexpensive gas laser, often home-built by hobbyists, which produces rather incoherent UV light at 337.1 nm. Metal ion lasers are gas lasers that generate deep ultraviolet wavelengths. Helium-silver (HeAg) 224 nm and neon-copper (NeCu) 248 nm are two examples. Like all low-pressure gas lasers, the gain media of these lasers have quite narrow oscillation linewidths, less than 3 GHz (0.5 picometers), making them candidates for use in fluorescence suppressed Raman spectroscopy.

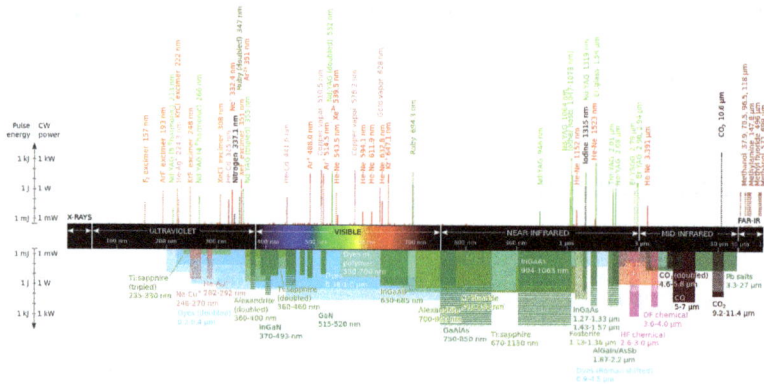

Wavelengths of commercially available lasers. Laser types with distinct laser lines are shown above the wavelength bar, while below are shown lasers that can emit in a wavelength range. The color codifies the type of laser material (see the figure description for more details).

Chemical Lasers

Chemical lasers are powered by a chemical reaction permitting a large amount of energy to be released quickly. Such very high power lasers are especially of interest to the military, however continuous wave chemical lasers at very high power levels, fed by streams of gasses, have been developed and have some industrial applications. As examples, in the hydrogen fluoride laser (2700–2900 nm) and the deuterium fluoride laser (3800 nm) the reaction is the combination of hydrogen or deuterium gas with combustion products of ethylene in nitrogen trifluoride.

Excimer Lasers

Excimer lasers are a special sort of gas laser powered by an electric discharge in which the lasing medium is an excimer, or more precisely an exciplex in existing designs. These are molecules which can only exist with one atom in an excited electronic state. Once the molecule transfers its excitation energy to a photon, therefore, its atoms are no longer bound to each other and the molecule disintegrates. This drastically reduces the population of the lower energy state thus greatly facilitating a population inversion. Excimers currently used are all noble gas compounds; noble gasses are chemically inert and can only form compounds while in an excited state. Excimer lasers typically operate at ultraviolet wavelengths with major applications including semiconductor photolithography and LASIK eye surgery. Commonly used excimer molecules include ArF (emission at 193 nm), KrCl (222 nm), KrF (248 nm), XeCl (308 nm), and XeF (351 nm). The molecular fluorine laser, emitting at 157 nm in the vacuum ultraviolet is sometimes referred to as an excimer laser, however this appears to be a misnomer inasmuch as F_2 is a stable compound.

Solid-State Lasers

A 50 W FASOR, based on a Nd:YAG laser, used at the Starfire Optical Range.

Solid-state lasers use a crystalline or glass rod which is "doped" with ions that provide the required energy states. For example, the first working laser was a ruby laser, made from ruby (chromium-doped corundum). The population inversion is actually maintained in the dopant. These materials are pumped optically using a shorter wavelength than the lasing wavelength, often from a flashtube or from another laser. The usage of the term "solid-state" in laser physics is narrower than in typical use. Semiconductor lasers (laser diodes) are typically *not* referred to as solid-state lasers.

Neodymium is a common dopant in various solid-state laser crystals, including yttrium orthovanadate (Nd:YVO$_4$), yttrium lithium fluoride (Nd:YLF) and yttrium aluminium garnet (Nd:YAG). All these lasers can produce high powers in the infrared spectrum at 1064 nm. They are used for cutting, welding and marking of metals and other materials, and also in spectroscopy and for pumping dye lasers. These lasers are also commonly frequency doubled, tripled or quadrupled to produce 532 nm (green, visible), 355 nm and 266 nm (UV) beams, respectively. Frequency-doubled diode-pumped solid-state (DPSS) lasers are used to make bright green laser pointers.

Ytterbium, holmium, thulium, and erbium are other common "dopants" in solid-state lasers. Ytterbium is used in crystals such as Yb:YAG, Yb:KGW, Yb:KYW, Yb:SYS, Yb:BOYS, Yb:CaF$_2$, typically operating around 1020–1050 nm. They are potentially very efficient and high powered due to a small quantum defect. Extremely high powers in ultrashort pulses can be achieved with Yb:YAG. Holmium-doped YAG crystals emit at 2097 nm and form an efficient laser operating at infrared wavelengths strongly absorbed by water-bearing tissues. The Ho-YAG is usually operated in a pulsed mode, and passed through optical fiber surgical devices to resurface joints, remove rot from teeth, vaporize cancers, and pulverize kidney and gall stones.

Titanium-doped sapphire (Ti:sapphire) produces a highly tunable infrared laser, commonly used for spectroscopy. It is also notable for use as a mode-locked laser producing ultrashort pulses of extremely high peak power.

Thermal limitations in solid-state lasers arise from unconverted pump power that heats the medium. This heat, when coupled with a high thermo-optic coefficient (dn/dT) can cause thermal lensing and reduce the quantum efficiency. Diode-pumped thin disk lasers overcome these issues by having a gain medium that is much thinner than the diameter of the pump beam. This allows for a more uniform temperature in the material. Thin disk lasers have been shown to produce beams of up to one kilowatt.

Fiber Lasers

Solid-state lasers or laser amplifiers where the light is guided due to the total internal reflection in a single mode optical fiber are instead called fiber lasers. Guiding of light allows extremely long gain regions providing good cooling conditions; fibers have high surface area to volume ratio which allows efficient cooling. In addition, the fiber's waveguiding properties tend to reduce thermal distortion of the beam. Erbium and ytterbium ions are common active species in such lasers.

Quite often, the fiber laser is designed as a double-clad fiber. This type of fiber consists of a fiber core, an inner cladding and an outer cladding. The index of the three concentric layers is chosen so that the fiber core acts as a single-mode fiber for the laser emission while the outer cladding acts as a highly multimode core for the pump laser. This lets the pump propagate a large amount of power into and through the active inner core region, while still having a high numerical aperture (NA) to have easy launching conditions.

Pump light can be used more efficiently by creating a fiber disk laser, or a stack of such lasers.

Fiber lasers have a fundamental limit in that the intensity of the light in the fiber cannot be so high that optical nonlinearities induced by the local electric field strength can become dominant and prevent laser operation and/or lead to the material destruction of the fiber. This effect is called photodarkening. In bulk laser materials, the cooling is not so efficient, and it is difficult to separate the effects of photodarkening from the thermal effects, but the experiments in fibers show that the photodarkening can be attributed to the formation of long-living color centers.

Photonic Crystal Lasers

Photonic crystal lasers are lasers based on nano-structures that provide the mode confinement and the density of optical states (DOS) structure required for the feedback to take place. They are typical micrometer-sized and tunable on the bands of the photonic crystals.

Semiconductor Lasers

A 5.6 mm 'closed can' commercial laser diode, probably from a CD or DVD player

Semiconductor lasers are diodes which are electrically pumped. Recombination of electrons and holes created by the applied current introduces optical gain. Reflection from the ends of the crystal form an optical resonator, although the resonator can be external to the semiconductor in some designs.

Commercial laser diodes emit at wavelengths from 375 nm to 3500 nm. Low to medium power laser diodes are used in laser pointers, laser printers and CD/DVD players.

Laser diodes are also frequently used to optically pump other lasers with high efficiency. The highest power industrial laser diodes, with power up to 10 kW (70 dBm), are used in industry for cutting and welding. External-cavity semiconductor lasers have a semiconductor active medium in a larger cavity. These devices can generate high power outputs with good beam quality, wavelength-tunable narrow-linewidth radiation, or ultrashort laser pulses.

In 2012, Nichia and OSRAM developed and manufactured commercial high-power green laser diodes (515/520 nm), which compete with traditional diode-pumped solid-state lasers.

Vertical cavity surface-emitting lasers (VCSELs) are semiconductor lasers whose emission direction is perpendicular to the surface of the wafer. VCSEL devices typically have a more circular output beam than conventional laser diodes. As of 2005, only 850 nm VCSELs are widely available, with 1300 nm VCSELs beginning to be commercialized, and 1550 nm devices an area of research. VECSELs are external-cavity VCSELs. Quantum cascade lasers are semiconductor lasers that have an active transition between energy *sub-bands* of an electron in a structure containing several quantum wells.

The development of a silicon laser is important in the field of optical computing. Silicon is the material of choice for integrated circuits, and so electronic and silicon photonic components (such as optical interconnects) could be fabricated on the same chip. Unfortunately, silicon is a difficult lasing material to deal with, since it has certain properties which block lasing. However, recently teams have produced silicon lasers through methods such as fabricating the lasing material from silicon and other semiconductor materials, such as indium(III) phosphide or gallium(III) arsenide, materials which allow coherent light to be produced from silicon. These are called hybrid silicon laser. Another type is a Raman laser, which takes advantage of Raman scattering to produce a laser from materials such as silicon.

Dye Lasers

Close-up of a table-top dye laser based on Rhodamine 6G

Dye lasers use an organic dye as the gain medium. The wide gain spectrum of available dyes, or mixtures of dyes, allows these lasers to be highly tunable, or to produce very short-duration pulses (on the order of a few femtoseconds). Although these tunable lasers are mainly known in their liquid form, researchers have also demonstrated narrow-linewidth tunable emission in dispersive oscillator configurations incorporating solid-state dye gain media. In their most prevalent form these solid state dye lasers use dye-doped polymers as laser media.

Free-Electron Lasers

The free-electron laser *FELIX* at the FOM Institute for Plasma Physics Rijnhuizen, Nieuwegein

Free-electron lasers, or FELs, generate coherent, high power radiation that is widely tunable, currently ranging in wavelength from microwaves through terahertz radiation and infrared to the visible spectrum, to soft X-rays. They have the widest frequency range of any laser type. While FEL beams share the same optical traits as other lasers, such as coherent radiation, FEL operation is quite different. Unlike gas, liquid, or solid-state lasers, which rely on bound atomic or molecular states, FELs use a relativistic electron beam as the lasing medium, hence the term *free-electron*.

Exotic Media

The pursuit of a high-quantum-energy laser using transitions between isomeric states of an atomic nucleus has been the subject of wide-ranging academic research since the early 1970s. Much of this is summarized in three review articles. This research has been international in scope, but mainly based in the former Soviet Union and the United States. While many scientists remain optimistic that a breakthrough is near, an operational gamma-ray laser is yet to be realized.

Some of the early studies were directed toward short pulses of neutrons exciting the upper isomer state in a solid so the gamma-ray transition could benefit from the line-narrowing of Mössbauer effect. In conjunction, several advantages were expected from two-stage pumping of a three-level system. It was conjectured that the nucleus of an atom, embedded in the near field of a laser-driven coherently-oscillating electron cloud would experience a larger dipole field than that of the driving laser. Furthermore, nonlinearity of the oscillating cloud would produce both spatial and temporal harmonics, so nuclear transitions of higher multipolarity could also be driven at multiples of the laser frequency.

In September 2007, the BBC News reported that there was speculation about the possibility of using positronium annihilation to drive a very powerful gamma ray laser. Dr. David Cassidy of the University of California, Riverside proposed that a single such laser could be used to ignite a nuclear fusion reaction, replacing the banks of hundreds of lasers currently employed in inertial confinement fusion experiments.

Space-based X-ray lasers pumped by a nuclear explosion have also been proposed as antimissile weapons. Such devices would be one-shot weapons.

Living cells have been used to produce laser light. The cells were genetically engineered to produce green fluorescent protein (GFP). The GFP is used as the laser's "gain medium", where light amplification takes place. The cells were then placed between two tiny mirrors, just 20 millionths of a meter across, which acted as the "laser cavity" in which light could bounce many times through the cell. Upon bathing the cell with blue light, it could be seen to emit directed and intense green laser light.

Uses

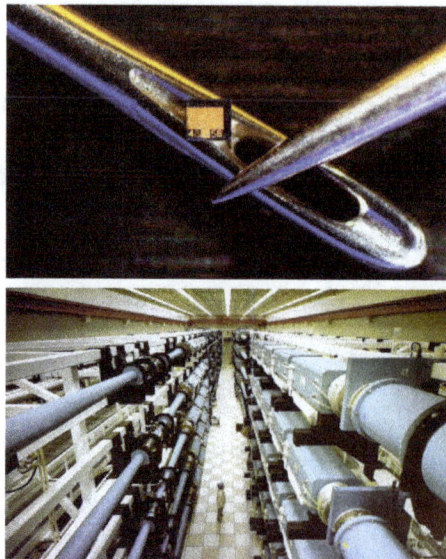

Lasers range in size from microscopic diode lasers (top) with numerous applications, to football field sized neodymium glass lasers (bottom) used for inertial confinement fusion, nuclear weapons research and other high energy density physics experiments.

When lasers were invented in 1960, they were called "a solution looking for a problem". Since then, they have become ubiquitous, finding utility in thousands of highly varied applications in every section of modern society, including consumer electronics, information technology, science, medicine, industry, law enforcement, entertainment, and the military. Fiber-optic communication using lasers is a key technology in modern communications, allowing services such as the Internet.

The first use of lasers in the daily lives of the general population was the supermarket barcode scanner, introduced in 1974. The laserdisc player, introduced in 1978, was the first successful consumer product to include a laser but the compact disc player was the first laser-equipped device to become common, beginning in 1982 followed shortly by laser printers.

Some other uses are:

- Medicine: Bloodless surgery, laser healing, surgical treatment, kidney stone treatment, eye treatment, dentistry.

- Industry: Cutting, welding, material heat treatment, marking parts, non-contact measurement of parts.

- Military: Marking targets, guiding munitions, missile defence, electro-optical countermeasures (EOCM), alternative to radar, blinding troops.

- Law enforcement: used for latent fingerprint detection in the forensic identification field

- Research: Spectroscopy, laser ablation, laser annealing, laser scattering, laser interferometry, lidar, laser capture microdissection, fluorescence microscopy, metrology.

- Product development/commercial: laser printers, optical discs (e.g. CDs and the like), barcode scanners, thermometers, laser pointers, holograms, bubblegrams.

- Laser lighting displays: Laser light shows.

- Cosmetic skin treatments: acne treatment, cellulite and striae reduction, and hair removal.

In 2004, excluding diode lasers, approximately 131,000 lasers were sold with a value of US$2.19 billion. In the same year, approximately 733 million diode lasers, valued at $3.20 billion, were sold.

Examples by Power

Different applications need lasers with different output powers. Lasers that produce a continuous beam or a series of short pulses can be compared on the basis of their average power. Lasers that produce pulses can also be characterized based on the *peak* power of each pulse. The peak power of a pulsed laser is many orders of magnitude greater than its average power. The average output power is always less than the power consumed.

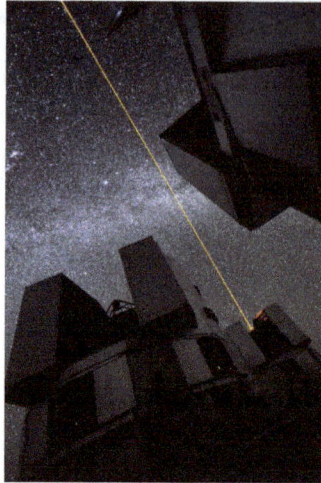

Laser application in astronomical adaptive optics imaging

The continuous or average power required for some uses:	
Power	Use
1–5 mW	Laser pointers
5 mW	CD-ROM drive
5–10 mW	DVD player or DVD-ROM drive
100 mW	High-speed CD-RW burner
250 mW	Consumer 16× DVD-R burner
400 mW	Burning through a jewel case including disc within 4 seconds
	DVD 24× dual-layer recording.
1 W	Green laser in current Holographic Versatile Disc prototype development
1–20 W	Output of the majority of commercially available solid-state lasers used for micro machining
30–100 W	Typical sealed CO_2 surgical lasers
100–3000 W	Typical sealed CO_2 lasers used in industrial laser cutting

Examples of pulsed systems with high peak power:

- 700 TW (700×10^{12} W) – National Ignition Facility, a 192-beam, 1.8-megajoule laser system adjoining a 10-meter-diameter target chamber.

- 1.3 PW (1.3×10^{15} W) – world's most powerful laser as of 1998, located at the Lawrence Livermore Laboratory

Hobby Uses

In recent years, some hobbyists have taken interests in lasers. Lasers used by hobbyists are generally of class IIIa or IIIb (see Safety), although some have made their own class

IV types. However, compared to other hobbyists, laser hobbyists are far less common, due to the cost and potential dangers involved. Due to the cost of lasers, some hobbyists use inexpensive means to obtain lasers, such as salvaging laser diodes from broken DVD players (red), Blu-ray players (violet), or even higher power laser diodes from CD or DVD burners.

Hobbyists also have been taking surplus pulsed lasers from retired military applications and modifying them for pulsed holography. Pulsed Ruby and pulsed YAG lasers have been used.

As Weapons

The US-Israeli Tactical High Energy weapon has been used to shoot down rockets and artillery shells.

Lasers of all but the lowest powers can potentially be used as incapacitating weapons, through their ability to produce temporary or permanent vision loss in varying degrees when aimed at the eyes. The degree, character, and duration of vision impairment caused by eye exposure to laser light varies with the power of the laser, the wavelength(s), the collimation of the beam, the exact orientation of the beam, and the duration of exposure. Lasers of even a fraction of a watt in power can produce immediate, permanent vision loss under certain conditions, making such lasers potential non-lethal but incapacitating weapons. The extreme handicap that laser-induced blindness represents makes the use of lasers even as non-lethal weapons morally controversial, and weapons designed to cause blindness have been banned by the Protocol on Blinding Laser Weapons. Incidents of pilots being exposed to lasers while flying have prompted aviation authorities to implement special procedures to deal with such hazards.

Laser weapons capable of directly damaging or destroying a target in combat are still in the experimental stage. The general idea of laser-beam weaponry is to hit a target with a train of brief pulses of light. The rapid evaporation and expansion of the surface causes shockwaves that damage the target. The power needed to project a high-powered laser beam of this kind is beyond the limit of current mobile power technology, thus favoring chemically powered gas dynamic lasers. Example experimental systems include MIRACL and the Tactical High Energy Laser.

Boeing YAL-1. The laser system is mounted in a turret attached to the aircraft nose

Throughout the 2000s, the United States Air Force worked on the Boeing YAL-1, an airborne laser mounted in a Boeing 747. It was intended to be used to shoot down incoming ballistic missiles over enemy territory. In March 2009, Northrop Grumman claimed that its engineers in Redondo Beach had successfully built and tested an electrically powered solid state laser capable of producing a 100-kilowatt beam, powerful enough to destroy an airplane. According to Brian Strickland, manager for the United States Army's Joint High Power Solid State Laser program, an electrically powered laser is capable of being mounted in an aircraft, ship, or other vehicle because it requires much less space for its supporting equipment than a chemical laser. However, the source of such a large electrical power in a mobile application remained unclear. Ultimately, the project was deemed to be infeasible, and was cancelled in December 2011, with the Boeing YAL-1 prototype being stored and eventually dismantled.

The United States Navy is developing a laser weapon referred to as the Laser Weapon System or LaWS.

Telecommunications in Space

Recent technology has allowed prototypes for laser communications and visible light communication in outer space. The communication range of free-space optical communication is currently of the order of several thousand kilometers, but has the potential to bridge interplanetary distances of millions of kilometers, using optical telescopes as beam expanders.

Safety

AVOID EXPOSURE
LASER RADIATION EMITTED
FROM THIS APERTURE

DANGER

LASER RADIATION
AVOID DIRECT EYE EXPOSURE

Max Output< 100mW
WAVELENGTH 532nm
Class IIIb Laser Product

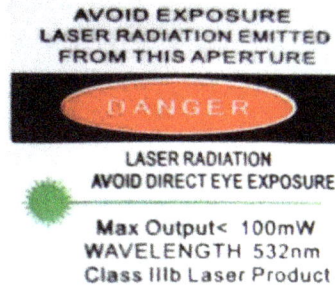

Left: European laser warning symbol required for Class 2 lasers and higher. Right: US laser warning label, in this case for a Class 3B laser

Even the first laser was recognized as being potentially dangerous. Theodore Maiman characterized the first laser as having a power of one "Gillette" as it could burn through one Gillette razor blade. Today, it is accepted that even low-power lasers with only a few milliwatts of output power can be hazardous to human eyesight when the beam hits the eye directly or after reflection from a shiny surface. At wavelengths which the cornea and the lens can focus well, the coherence and low divergence of laser light means that it can be focused by the eye into an extremely small spot on the retina, resulting in localized burning and permanent damage in seconds or even less time.

Lasers are usually labeled with a safety class number, which identifies how dangerous the laser is:

- Class 1 is inherently safe, usually because the light is contained in an enclosure, for example in CD players.

- Class 2 is safe during normal use; the blink reflex of the eye will prevent damage. Usually up to 1 mW power, for example laser pointers.

- Class 3R (formerly IIIa) lasers are usually up to 5 mW and involve a small risk of eye damage within the time of the blink reflex. Staring into such a beam for several seconds is likely to cause damage to a spot on the retina.

- Class 3B can cause immediate eye damage upon exposure.

- Class 4 lasers can burn skin, and in some cases, even scattered light can cause eye and/or skin damage. Many industrial and scientific lasers are in this class.

The indicated powers are for visible-light, continuous-wave lasers. For pulsed lasers and invisible wavelengths, other power limits apply. People working with class 3B and class 4 lasers can protect their eyes with safety goggles which are designed to absorb light of a particular wavelength.

Infrared lasers with wavelengths longer than about 1.4 micrometers are often referred to as "eye-safe", because the cornea tends to absorb light at these wavelengths, protecting the retina from damage. The label "eye-safe" can be misleading, however, as it applies only to relatively low power continuous wave beams; a high power or Q-switched laser at these wavelengths can burn the cornea, causing severe eye damage, and even moderate power lasers can injure the eye.

References

- Chu, Steven; Townes, Charles (2003). "Arthur Schawlow". In Edward P. Lazear (ed.),. Biographical Memoirs. vol. 83. National Academy of Sciences. p. 202. ISBN 0-309-08699-X.

- Bertolotti, Mario (2015), Masers and Lasers, Second Edition: An Historical Approach, CRC Press, pp. 89–91, ISBN 9781482217803, retrieved March 15, 2016

- Townes, Charles H. (1999). How the Laser Happened: Adventures of a Scientist, Oxford University Press, ISBN 9780195122688.

- Charles H. Townes (2003). "The first laser". In Laura Garwin; Tim Lincoln. A Century of Nature: Twenty-One Discoveries that Changed Science and the World. University of Chicago Press. pp. 107–12. ISBN 0-226-28413-1. Retrieved February 2, 2008.

- Maks, Stephanie. "Howto: Make a DVD burner into a high-powered laser". Transmissions from Planet Stephanie. Retrieved April 6, 2015.

- Luis Martinez (9 Apr 2013). "Navy's New Laser Weapon Blasts Bad Guys From Air, Sea". ABC. Retrieved 9 April 2013.

- Boroson, Don M. (2005), Optical Communications: A Compendium of Signal Formats, Receiver Architectures, Analysis Mathematics, and Performance Characteristics, retrieved 8 Jan 2013

- "Another world first for Artemis: a laser link with an aircraft". European Space Agency. December 18, 2006. Retrieved June 28, 2011.

- Malte C. Gather & Seok Hyun Yun (June 12, 2011). "Single-cell biological lasers". Nature Photonics. Retrieved June 13, 2011.

2

Types of Laser

Lasers can be classified based on the gain medium used to amplify light. They can be classified as gas laser, chemical laser, dye laser, solid-state laser, gas dynamic laser, free-electron laser, Raman laser, nuclear pumped laser, excimer laser, fiber laser etc. Gas laser can further be divided into nitrogen laser, helium-neon laser and ion laser. This chapter describes each of these lasers in detail explaining their discerning features and working mechanisms.

Gas Laser

A gas laser is a laser in which an electric current is discharged through a gas to produce coherent light. The gas laser was the first continuous-light laser and the first laser to operate on the principle of converting electrical energy to a laser light output. The first gas laser, the Helium–neon laser (HeNe), was co-invented by Iranian-Azerbaijani physicist Ali Javan and American physicist William R. Bennett, Jr. in 1960. It produced a coherent light beam in the infrared region of the spectrum at 1.15 micrometres.

Types of Gas Laser

Gas lasers using many gases have been built and used for many purposes.

Carbon dioxide lasers, or CO_2 lasers can emit hundreds of kilowatts at 9.6 μm and 10.6 μm, and are often used in industry for cutting and welding. The efficiency of a CO_2 laser is over 10%.

Carbon monoxide or "CO" lasers have the potential for very large outputs, but the use of this type of laser is limited by the toxicity of carbon monoxide gas. Human operators must be protected from this deadly gas. Furthermore, it is extremely corrosive to many materials including seals, gaskets, etc.

Helium–neon (HeNe) lasers can be made to oscillate at over 160 different wavelengths by adjusting the cavity Q to peak at the desired wavelength. This can be done by adjusting the spectral response of the mirrors or by using a dispersive element (Littrow prism) in the cavity. Units operating at 633 nm are very common in schools and laboratories because of their low cost and near perfect beam qualities.

Nitrogen lasers operate in the ultraviolet range, typically 337.1 nm, using molecular nitrogen as its gain medium, pumped by an electrical discharge.

TEA lasers are energized by a high voltage electrical discharge in a gas mixture generally at or above atmospheric pressure. The acronym "TEA" stands for Transversely Excited Atmospheric.

Chemical Lasers

Chemical lasers are powered by a chemical reaction, and can achieve high powers in continuous operation. For example, in the hydrogen fluoride laser (2700–2900 nm) and the deuterium fluoride laser (3800 nm) the reaction is the combination of hydrogen or deuterium gas with combustion products of ethylene in nitrogen trifluoride. They were invented by George C. Pimentel.

Chemical lasers are powered by a chemical reaction permitting a large amount of energy to be released quickly. Such very high power lasers are especially of interest to the military, however continuous wave chemical lasers at very high power levels, fed by streams of gasses, have been developed and have some industrial applications. As examples, in the hydrogen fluoride laser (2700–2900 nm) and the deuterium fluoride laser (3800 nm) the reaction is the combination of hydrogen or deuterium gas with combustion products of ethylene in nitrogen trifluoride

Excimer Lasers

Excimer lasers are powered by a chemical reaction involving an *excited dimer*, or *excimer*, which is a short-lived dimeric or heterodimeric molecule formed from two species (atoms), at least one of which is in an excited electronic state. They typically produce ultraviolet light, and are used in semiconductor photolithography and in LASIK eye surgery. Commonly used excimer molecules include F_2 (fluorine, emitting at 157 nm), and noble gas compounds (ArF [193 nm], KrCl [222 nm], KrF [248 nm], XeCl [308 nm], and XeF [351 nm]).

Ion Lasers

Argon-ion lasers emit light in the range 351–528.7 nm. Depending on the optics and the laser tube a different number of lines is usable but the most commonly used lines are 458 nm, 488 nm and 514.5 nm.

Metal-Vapor Lasers

Metal-vapor lasers are gas lasers that typically generate ultraviolet wavelengths. Helium-silver (HeAg) 224 nm neon-copper (NeCu) 248 nm and helium-cadmium (HeCd) 325 nm are three examples. These lasers have particularly narrow oscillation linewidths of less than 3 GHz (0.5 picometers), making them candidates for use in fluorescence suppressed Raman spectroscopy.

The Copper vapor laser, with two spectral lines of green (510.6 nm) and yellow

(578.2 nm), is the most powerful laser with the highest efficiency in the visible spectrum.

Advantages

- High volume of active material
- Active material is relatively inexpensive
- Almost impossible to damage the active material
- Heat can be removed quickly from the cavity

Applications

- He–Ne laser is mainly used in making holograms.
- In laser printing He–Ne laser is used as a source for writing on the photosensitive material.
- He–Ne lasers were used in reading the Bar Code which is imprinted on the product. They have been largely replaced by laser diodes.
- Nitrogen lasers and excimer laser are used in pulsed dye laser pumping.
- Ion lasers, mostly argon, are used in CW dye laser pumping.

Various Types of Gas Laser

Nitrogen laser

A 337nm wavelength and 170 μJ pulse energy 20 Hz cartridge nitrogen laser

A nitrogen laser is a gas laser operating in the ultraviolet range (typically 337.1 nm) using molecular nitrogen as its gain medium, pumped by an electrical discharge.

The wall-plug efficiency of the nitrogen laser is low, typically 0.1% or less, though nitrogen lasers with efficiency of up to 3% have been reported in the literature. The wall-plug efficiency is the product of the following three efficiencies:

- electrical: TEA laser

- gain medium: This is the same for all nitrogen lasers and thus has to be at least 3%

  - inversion by electron impact is 10 to 1 due to Franck–Condon principle

  - energy lost in the lower laser level: 40%

- optical: More induced emission than spontaneous emission

Gain Medium

The gain medium is nitrogen molecules in the gas phase. The nitrogen laser is a three-level laser. In contrast to more typical four-level lasers, the upper laser level of nitrogen is directly pumped, imposing no speed limits on the pump. Pumping is normally provided by direct electron impact; the electrons must have sufficient energy, or they will fail to excite the upper laser level. Typically reported optimum values are in the range of 80 to 100 eV per Torr·cm pressure of nitrogen gas.

There is a 40 ns upper limit of laser lifetime at low pressures and the lifetime becomes shorter as the pressure increases. The lifetime is only 1 to 2 ns at 1 atmosphere. In general

$$t[\text{ns}] = \frac{36}{1 + 12.8 * p[\text{bar}]}$$

The strongest lines are at 337.1 nm wavelength in the ultraviolet. Other lines have been reported at 357.6 nm, also ultraviolet. This information refers to the second positive system of molecular nitrogen, which is by far the most common. No vibration of the two nitrogen atoms is involved, because the atom-atom distance does not change with the electronic transition. The rotation needs to change to deliver the angular momentum of the photon, furthermore multiple rotational states are populated at room temperature. There are also lines in the far-red and infrared from the first positive system, and a visible blue laser line from the molecular nitrogen positive (1+) ion.

The metastable lower level lifetime is 40 µs, thus, the laser self-terminates, typically in less than 20 ns. This type of self-termination is sometimes referred to as "bottlenecking in the lower level". This is only a rule of thumb as is seen in many other lasers: The helium–neon laser also has a bottleneck as one decay step needs the walls of the cavity and this laser typically runs in continuous mode. Several organic dyes with upper level lifetimes of less than 10 ns have been used in continuous mode. The Nd:YAG laser has an upper level lifetime of 230 µs, yet it also supports 100 ps pulses.

Repetition rates can range as high as a few kHz, provided adequate gas flow and cooling of the structure are provided. Cold nitrogen is a better medium than hot nitrogen, and this appears to be part of the reason that the pulse energy and power drop as the repetition rate increases to more than a few pulses per second. There are also, apparently, issues involving ions remaining in the laser channel.

Air, which is 78% nitrogen, can be used, but more than 0.5% oxygen poisons the laser.

Optics

Nitrogen lasers can operate within a resonator cavity, but due to the typical gain of 2 every 20 mm they more often operate on superluminescence alone; though it is common to put a mirror at one end such that the output is emitted from the opposite end.

For a 10 mm wide gain volume diffraction comes into play after 30 m along the gain medium, a length which is unheard of. Thus this laser does not need a concave lens or refocusing lenses and beam quality improves along the gain medium. The height of the pumped volume may be as small as 1 mm, needing a refocusing lens already after 0.3 m. A simple solution is to use rounded electrodes with a large radius, so that a quadratic pump profile is obtained.

Electrical

The gain medium is usually pumped by a transverse electrical discharge. When the pressure is at (or above) 1013 mbar (atmospheric pressure), the configuration is called a TEA laser **T**ransverse **E**lectrical discharge in gas at **A**tmospheric pressure, this is also used for pressures down to 30 mbar.

Microscopic Description of a Fast Discharge

In a strong external electric field this electron creates an electron avalanche in the direction of the electric field lines. Diffusion of electrons and elastic scattering at a buffer gas molecule spreads the avalanche perpendicular to the field. Inelastic scattering creates photons, which create new avalanches centimeters away. After some time the electric charge in the avalanche becomes so large that following Coulomb's law it generates an electric field as large as the external electric field. At regions of increased field strength the avalanche effect is enhanced. This leads to electric arc like discharges called streamers. A mix of a noble gas (up to 0.9) and nitrogen enhance elastic scattering of electrons over electron multiplying and thus widens avalanches and streamers.

Spark gaps use a high density of gas molecules and a low density of initial electrons to favor streamers. Electrons are removed by a slowly rising voltage. A high density gas increases the breakdown field, thus shorter arcs can be used with lower inductance and the capacity between the electrodes is increased. A wide streamer has a lower inductance.

Gas lasers use low density of gas molecules and a high density of initial electrons to prevent streamers. Electrons are added by preionisation not removed by oxygen, because nitrogen from bottles is used. Wide avalanches can excite more nitrogen molecules.

Inelastic scattering heats up a molecule, so that in a second scattering the probability of electron emission is increased. This leads to an arc. Typically arcing occurs *after* lasing in nitrogen. The streamer in the spark gap discharges the electrodes only by means of image charge, thus when the streamer touches both electrodes most of the charge is still available to feed the arc, additional charge is stored on the distribution plates. Thus arcing in the spark gap starts *before* lasing.

Conditions for pulsed avalanche discharges are described by Levatter and Lin.

Electrodynamics

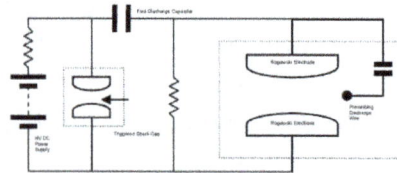

Circuit.

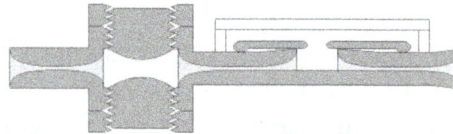

Low inductance implementation cross cut. Erratum: Right cap needs to be bigger.

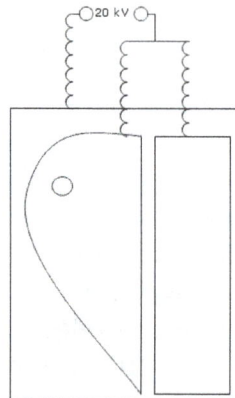

Low inductance implementation top view. Erratum: Caps should be slightly longer than the channel and have rounded corners.

The electronics is a circuit composed of a spark gap, a capacitor, and the discharge through the nitrogen. First the spark gap and the capacitor are charged. The spark gap then discharges itself and voltage is applied to the nitrogen.

An alternative construction uses two capacitors connected as a Blumlein generator. Two capacitors are connected so that one plate is a common earth, the others are each connected to the spark gap electrodes. These capacitors are often constructed from a single layer of printed circuit board, or similar stack of copper foil and thin dielectric.

The capacitors are linked through an inductor, a simple air-spaced coil. One capacitor also has a small spark gap across it. When HT is applied, the two capacitors are charged slowly, effectively linked by the inductor. When the spark gap reaches its triggering voltage, it discharges and quickly reduces that capacitor's voltage to zero. As the discharge is rapid, the inductor acts as an open circuit and so the voltage difference across the transverse spark gap (between the two capacitors) rises rapidly until the main spark gap discharges, firing the laser.

The speed of either circuit is increased in two steps. First, the inductance of all components is reduced by shortening and widening conductors and by squeezing the circuit into a flat rectangle. The total inductance is the sum of the components:

object	length	thickness	width	width	inductance	inductance	inductance	capacity	oscillation
			as coil	as wire	measured	coil theory	wire theory	plate theory	period
unit	m	m	m	m	nH	nH	nH	nF	ns
spark gap	2×10^{-2}	1×10^{-2}	2×10^{-2}	1×10^{-5}	10	12.57	13.70	0.0004	
metal tape	2×10^{-2}	2×10^{-2}	4×10^{-2}	5×10^{-3}		12.57	5.32	0.0004	
cap. 1	2×10^{-1}	4×10^{-4}	3×10^{-1}			0.34		2.6563	
metal tape	2×10^{-2}	2×10^{-2}	3×10^{-1}			1.68		0.0027	
laser channel	1×10^{-2}	2×10^{-2}	3×10^{-1}			0.84		0.0013	
metal tape	2×10^{-2}	2×10^{-2}	3×10^{-1}			1.68		0.0027	
cap. 2	3×10^{-1}	4×10^{-4}	3×10^{-1}			0.50		3.9844	
spark osc.						22.90		2.6563	49
disch. osc.						5.03		1.5938	18

The intense discharge is reported to distort oscilloscopes nearby. This can be reduced by building the laser symmetrically into a grounded cylinder with the spark gap at the bottom, the laser at the top, capacitor 1 left and right, and capacitor 2 left and right stacked onto capacitor 1. This has the further advantage of reducing the inductance. And this has the disadvantage that the laser channel cannot be inspected for sparks anymore.

Secondly, transmission line theory and waveguide theory is applied to achieve a traveling wave excitation. Measured nitrogen laser pulses are so long that the second step is unimportant. From this analysis it follows that:

- the end mirror and the spark gap are on the same side

- a long narrow laser at atmospheric pressures is ineffective

Spark Gap

Paschen's law states that the length of the spark gap is inverse-proportional to the pressure. For a fixed length to diameter ratio of the spark, the inductance is proportional to the length (source , compare with: dipole antenna). Thus the electrodes of the spark gap are glued or welded on a dielectric spacer-ring. To reduce the danger due to the pressure, the volume is minimized. To prevent sparks outside space ring in the low pressure the spacer usually gets thicker outwards in an s-shaped manner.

Connection between spark gap and laser channel based on traveling wave theory:

- The low inductance spark gap may be inserted into a strip transmission line

- biconical spark gap

- biconical spark gap

- biconical spark gap

The breakdown voltage is low for helium, medium for nitrogen and high for SF_6, though nothing is said about the spark thickness variations.

8E10A/s are possible with a spark gap this nicely matches the typical rise times of 1E-8s and typical currents of 1E3A occurring in nitrogen lasers.

A cascade of spark gaps allows to use a weak trigger pulse to initiate a streamer in the smaller gap, wait for its transition into an arc, and then for this arc to extend into the larger gap. Still the first spark gap in the cascade needs a free electron to start with, so jitter is rather high.

Preionisation

Avalanches homogenize a discharge fast mostly along the field lines. With a short duration (<10 ms) since the last laser pulse enough ions are left over so that all avalanches overlap also laterally. With low pressure (<100 kPa) the max charge carrier density is low and the electromagnetic driven transition from avalanche to spark is inhibited.

In other cases UV radiation homogenizes a discharge slowly perpendicular to a discharge. These are brought into balance by placing two linear discharges next to each other 1 cm apart. The first discharge is across a smaller gap and starts early. Due to the low number of initial electrons streamers typically 1 mm apart are seen. The electrodes for the first discharge are covered by a dielectric, which limits this discharge. Therefore the voltage is able to rise further until avalanches can start in the second gap. These are so many that they overlap and excite every molecule.

With about 11 ns the UV generation, ionisation, and electron capture are in a similar speed regime as the nitrogen laser pulse duration and thus as fast electric must be applied.

Excitation by Electron Impact

The upper laser level is excited efficiently by electrons with more than 11 eV, best energy is 15 eV. The electron temperature in the streamers only reaches 0.7 eV. Helium due to its higher ionisation energy and lack of vibrational excitations increases the temperature to 2.2 eV. Higher voltages increase the temperature. Higher voltages mean shorter pulses.

Typical Devices

The gas pressure in a nitrogen laser ranges from a few mbar to as much as several bar. Furthermore, air provides significantly less output energy than pure nitrogen or a mixture of nitrogen and helium. The pulse energy ranges from μJ to mJ (a commercial version using a spark gap delivers 300 microJ) and a peak power in the range of kW to more than 3 MW can be achieved. The pulse temporal width is between a few hundred picoseconds (typically at 1 atmosphere partial pressure of nitrogen) and a maximum of approximately 30 nanoseconds at reduced pressure (typically some dozens of Torr), though fwhm pulsewidths of 6 to 8 ns are typical.

Amateur Construction

The transverse discharge nitrogen laser has long been a popular choice for amateur home construction, owing to its simple construction and simple gas handling. It was described by *Scientific American* in 1974, as one of the first laser home-construction articles.

Helium–Neon Laser

A helium–neon laser or HeNe laser, is a type of gas laser whose gain medium consists of a mixture of helium and neon(10:1) inside of a small bore capillary tube, usually excited by a DC electrical discharge. The best-known and most widely used HeNe laser operates at a wavelength of 632.8 nm, in the red part of the visible spectrum.

History of HeNe Laser Development

The first HeNe lasers emitted light at 1.15 μm, in the infrared spectrum, and were the first gas lasers. However, a laser that operated at visible wavelengths was much more in demand, and a number of other neon transitions were investigated to identify ones in which a population inversion can be achieved. The 633 nm line was found to have the highest gain in the visible spectrum, making this the wavelength of choice for most

HeNe lasers. However other visible as well as infrared stimulated emission wavelengths are possible, and by using mirror coatings with their peak reflectance at these other wavelengths, HeNe lasers could be engineered to employ those transitions; this includes visible lasers appearing red, orange, yellow, and green. Stimulated emissions are known from over 100 μm in the far infrared to 540 nm in the visible. Because visible transitions have somewhat lower gain, these lasers generally have lower output efficiencies and are more costly. The 3.39 μm transition has a very high gain but is prevented from use in an ordinary HeNe laser (of a different intended wavelength) because the cavity and mirrors are lossy at that wavelength. However in high power HeNe lasers having a particularly long cavity, superluminescence at 3.39 μm can become a nuisance, robbing power from the stimulated emission medium, often requiring additional suppression. The best-known and most widely used HeNe laser operates at a wavelength of 632.8 nm, in the red part of the visible spectrum. It was developed at Bell Telephone Laboratories in 1962, 18 months after the pioneering demonstration at the same laboratory of the first continuous infrared HeNe gas laser in December 1960.

Construction and Operation

The gain medium of the laser, as suggested by its name, is a mixture of helium and neon gases, in approximately a 10:1 ratio, contained at low pressure in a glass envelope. The gas mixture is mostly helium, so that helium atoms can be excited. The excited helium atoms collide with neon atoms, exciting some of them to the state that radiates 632.8 nm. Without helium, the neon atoms would be excited mostly to lower excited states responsible for non-laser lines. A neon laser with no helium can be constructed but it is much more difficult without this means of energy coupling. Therefore, a HeNe laser that has lost enough of its helium (e.g., due to diffusion through the seals or glass) will lose its laser functionality because the pumping efficiency will be too low. The energy or pump source of the laser is provided by a high voltage electrical discharge passed through the gas between electrodes (anode and cathode) within the tube. A DC current of 3 to 20 mA is typically required for CW operation. The optical cavity of the laser usually consists of two concave mirrors or one plane and one concave mirror, one having very high (typically 99.9%) reflectance and the output coupler mirror allowing approximately 1% transmission.

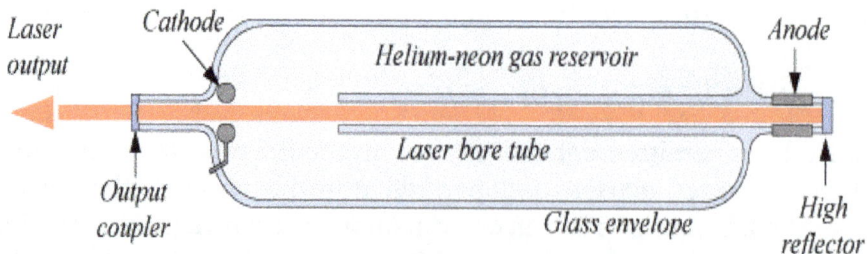

Schematic diagram of a helium–neon laser

Commercial HeNe lasers are relatively small devices, among gas lasers, having cavity lengths usually ranging from 15 cm to 50 cm (but sometimes up to about 1 metre to achieve the highest powers), and optical output power levels ranging from 0.5 to 50 mW.

The red HeNe laser wavelength of 633 nm has an actual vacuum wavelength of 632.991 nm, or about 632.816 nm in air. The wavelengths of the stimulated emission modes lie within about 0.001 nm above or below this value, and the wavelengths of those modes shift within this range due to thermal expansion and contraction of the cavity. Frequency-stabilized versions enable the wavelength of a single mode to be specified to within 1 part in 10^8 by the technique of comparing the powers of two longitudinal modes in opposite polarizations. Absolute stabilization of the laser's frequency (or wavelength) as fine as 2.5 parts in 10^{11} can be obtained through use of an iodine absorption cell.

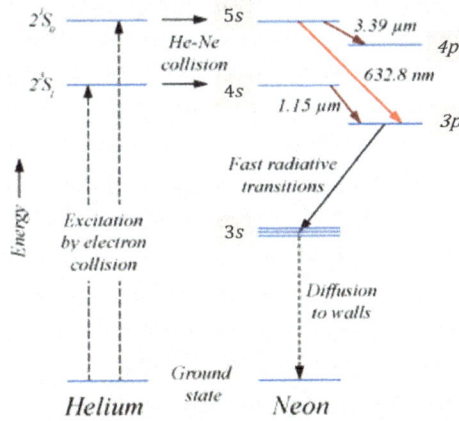

The mechanism producing population inversion and light amplification in a HeNe laser plasma originates with inelastic collision of energetic electrons with ground state helium atoms in the gas mixture. As shown in the accompanying energy level diagram, these collisions excite helium atoms from the ground state to higher energy excited states, among them the 2^3S_1 and 2^1S_0 long-lived metastable states. Because of a fortuitous near coincidence between the energy levels of the two He metastable states, and the $5s_2$ and $4s_2$ (Paschen notation) levels of neon, collisions between these helium metastable atoms and ground state neon atoms results in a selective and efficient transfer of excitation energy from the helium to neon. This excitation energy transfer process is given by the reaction equations:

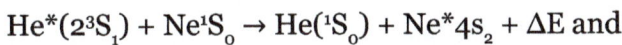

$$He^*(2^3S_1) + Ne^1S_0 \rightarrow He(^1S_0) + Ne^*4s_2 + \Delta E \text{ and}$$

$$He^*(2^1S) + Ne^1S_0 + \Delta E \rightarrow He(^1S_0) + Ne^*5s_2$$

where (*) represents an excited state, and ΔE is the small energy difference between the energy states of the two atoms, of the order of 0.05 eV or 387 cm^{-1}, which is supplied by kinetic energy. Excitation energy transfer increases the population of the neon $4s_2$ and $5s_2$ levels manyfold. When the population of these two upper levels exceeds that of the corresponding lower level neon state, $3p_4$ to which they are optically connected, population inversion is present. The medium becomes capable of amplifying light in a narrow band at 1.15 µm (corresponding to the $4s_2$ to $3p_4$ transition) and in a narrow band at 632.8 nm (corresponding to the $5s_2$ to $3p_4$ transition at 632.8 nm). The $3p_4$ level is efficiently emptied by fast radiative decay to the 1s state, eventually reaching the ground state.

The remaining step in utilizing optical amplification to create an optical oscillator is to place highly reflecting mirrors at each end of the amplifying medium so that a wave in a particular spatial mode will reflect back upon itself, gaining more power in each pass than is lost due to transmission through the mirrors and diffraction. When these conditions are met for one or more longitudinal modes then radiation in those modes will rapidly build up until gain saturation occurs, resulting in a stable continuous laser beam output through the front (typically 99% reflecting) mirror.

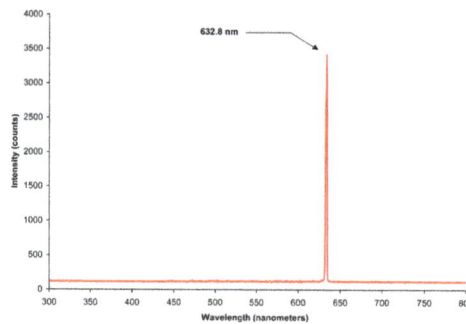

Spectrum of a helium neon laser illustrating its very high spectral purity (limited by the measuring apparatus). The 0.002 nm bandwidth of the stimulated emission medium is well over 10,000 times narrower than the spectral width of a light-emitting diode (whose spectrum is shown here for comparison), with the bandwidth of a single longi-tudinal mode being much narrower still.

The gain bandwidth of the HeNe laser is dominated by Doppler broadening rather than pressure broadening due to the low gas pressure, and is thus quite narrow: only about 1.5 GHz full width for the 633 nm transition. With cavities having typical lengths of 15 cm to 50 cm, this allows about 2 to 8 longitudinal modes to oscillate simultaneously (however single longitudinal mode units are available for special applications). The visible output of the red HeNe laser, long coherence length, and its excellent spatial quality, makes this laser a useful source for holography and as a wavelength reference for spectroscopy. A stabilized HeNe laser is also one of the benchmark systems for the definition of the meter.

Prior to the invention of cheap, abundant diode lasers, red HeNe lasers were widely used in barcode scanners at supermarket checkout counters. Laser gyroscopes have employed HeNe lasers operating at 0.633 μm in a ring laser configuration. HeNe lasers are generally present in educational and research optical laboratories.

Applications

Red HeNe lasers have many industrial and scientific uses. They are widely used in laboratory demonstrations in the field of optics because of their relatively low cost and ease of operation compared to other visible lasers producing beams of similar quality in terms of spatial coherence (a single-mode Gaussian beam) and long coherence length (however since about 1990 semiconductor lasers have offered a lower-cost alternative for many such applications). A consumer application of the red HeNe laser is the LaserDisc player, made by Pioneer. The laser is used in the device to read the optical disk.

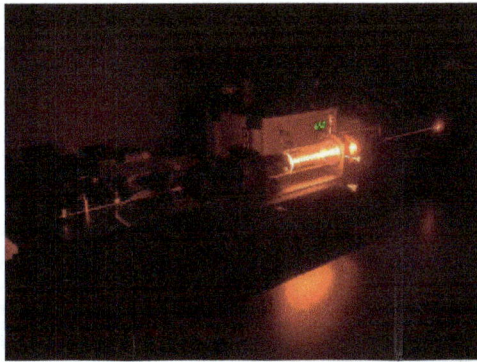

Helium-Neon Laser at the University of Chemnitz, Germany

Ion Laser

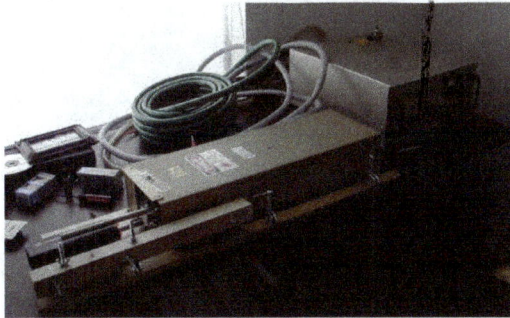

1 mW Uniphase HeNe on alignment rig (left) and 2 W Lexel 88 argon-ion laser (center) with power-supply (right). To the rear are hoses for water cooling.

An ion laser is a gas laser that uses an ionized gas as its lasing medium. Like other gas lasers, ion lasers feature a sealed cavity containing the laser medium and mirrors forming a Fabry–Pérot resonator. Unlike helium–neon lasers, the energy level transitions that contribute to laser action come from ions. Because of the large amount of energy required to excite the ionic transitions used in ion lasers, the required current is much greater, and as a result all but the smallest ion lasers are water-cooled. A small air-cooled ion laser might produce, for example, 130 mW of light with a tube current of 10 A at 105 V. This is a total power draw over 1 kW, which translates into a large amount of heat that must be dissipated.

Types

Krypton Laser

A krypton laser is an ion laser using krypton ions as a gain medium, pumped by electric discharge. Krypton lasers are used for scientific research, or when krypton is mixed with argon, for creation of "white-light" lasers, useful for laser light shows. Krypton lasers are also used in medicine (e.g. for coagulation of retina), for manufacture of security holograms, and numerous other purposes.

Krypton lasers emit at several wavelengths through the visible spectrum: at 406.7 nm, 413.1 nm, 415.4 nm, 468.0 nm, 476.2 nm, 482.5 nm, 520.8 nm, 530.9 nm, 568.2 nm, 647.1 nm, 676.4 nm.

Argon Laser

The argon-ion laser was invented in 1964 by William Bridges at Hughes Aircraft and is one of a family of ion lasers that use a noble gas as the active medium.

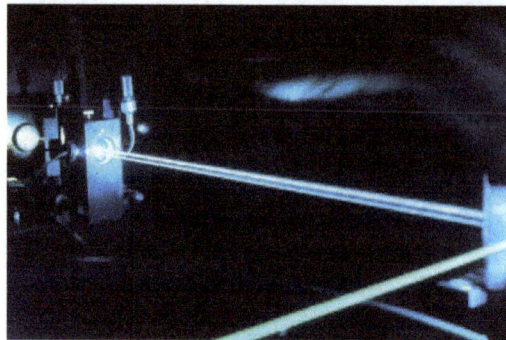

This argon-ion laser emits blue-green light at 488 and 514 nm

Argon-ion lasers are used for retinal phototherapy (for diabetes), lithography, and pumping other lasers. Argon-ion lasers emit at 13 wavelengths through the visible, ultraviolet, and near-visible spectrum, including: 351.1 nm, 363.8 nm, 454.6 nm, 457.9 nm, 465.8 nm, 476.5 nm, 488.0 nm, 496.5 nm, 501.7 nm, 514.5 nm, 528.7 nm, 1092.3 nm.

Common argon and krypton lasers are capable of emitting continuous-wave (CW) output of several milliwatts to tens of watts. Their tubes are usually made from nickel end bells, kovar metal-to-ceramic seals, beryllium oxide ceramics, or tungsten disks mounted on a copper heat spreader in a ceramic liner. The earliest tubes were simple quartz, followed by quartz with graphite disks. In comparison with the helium–neon lasers that require just a few milliamperes, the current used for pumping the krypton laser is several amperes, as the gas has to be ionized. The ion laser tube produces a lot of waste heat and requires active cooling.

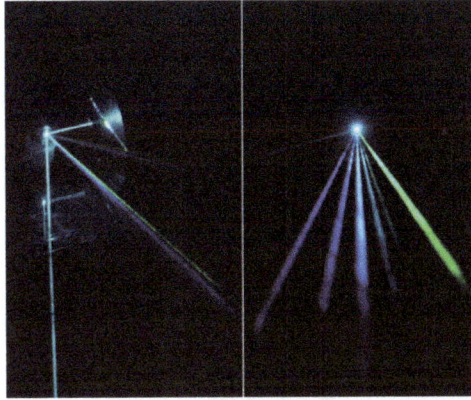

An argon-laser beam consisting of multiple colors (wavelengths) strikes a silicon diffraction mirror grating and is separated into several beams, one for each wavelength (left to right): 458 nm, 476 nm, 488 nm, 497 nm, 502 nm, 515 nm

The typical noble-gas ion-laser plasma consists of a high-current-density glow discharge in a noble gas in the presence of a magnetic field. Typical CW plasma conditions are current densities of 100 to 2000 A/cm^2, tube diameters of 1 to 10 mm, filling pressures of 0.1 to 1.0 Torr (0.0019 to 0.019 psi), and an axial magnetic field of the order of 1000 G.

William R. Bennett, the co-inventor of the first gas laser (the helium–neon laser), was the first to observe spectral hole burning effects in gas lasers and created a theory of hole burning effects on laser oscillation. He was co-discoverer of lasers using electron-impact excitation in each of the noble gases, dissociative excitation transfer in the neon–oxygen laser (the first chemical laser), and collision excitation in several metal-vapor lasers.

Other Commercially Available Types

- Ar/Kr: A mix of argon and krypton can result in a laser with output wavelengths that appear as white light.

- Helium–cadmium: blue laser emission at 442 nm and ultraviolet at 325 nm.

- Copper vapor: yellow and green emission at 578 nm and 510 nm.

Experimental

- Xenon

- Iodine

- Oxygen

Power Supplies

- NPN passbank like the Spectra-physics 270 supply
- MOSFET switchers like the Omnichrome 150 supply
- Early switchers used NPN_PNP Pairs, (i.e. American Laser or HGM Medical)
- Insulated-gate bipolar transistor (IGBT)units will be seen more in days to come
- Switched resistor (Spectra Physics)
- Non-switched resistor (Home-made, typically a water heater element)
- Water-cooled resistor (Laser Ionics etc.)
- Phased SCR power supplies similar to long xenon arc lamps are used in medical lasers to reduce expense (Coherent)
- Power on Demand power supplies are used for pulsed medical ion laser systems, these power supplies consist of a large capacitor bank charged by a switching supply to enable multi watt lasers to run off common single phase power supplies in doctor's offices.
- A typical air-cooled Argon Tube needs an equivalent series resistance of ~6 Ohms when running @ 10 amps off 117V power. The plasma in an ion laser, unlike a Helium Neon Laser, has a slightly positive resistance, but will still run away without ballasting. This is why ion laser supplies are very difficult to design. On a large frame laser, the plasma itself has an effective resistance of about -7 Ohms (Spectra Physics 171 Service Manual)

Chemical Laser

A chemical laser is a laser that obtains its energy from a chemical reaction. Chemical lasers can reach continuous wave output with power reaching to megawatt levels. They are used in industry for cutting and drilling.

Common examples of chemical lasers are the chemical oxygen iodine laser (COIL), all gas-phase iodine laser (AGIL), and the hydrogen fluoride (HF) and deuterium fluoride (DF) lasers, both operating in the mid-infrared region. There is also a DF–CO_2 laser (deuterium fluoride–carbon dioxide), which, like COIL, is a "transfer laser." The HF and DF lasers are unusual, in that there are several molecular energy transitions with sufficient energy to cross the threshold required for lasing. Since the molecules do not collide frequently enough to re-distribute the energy, several of these laser modes operate either simultaneously, or in extremely rapid succession, so that an HF or DF laser appears to operate simultaneously on several wavelengths unless a wavelength selection device is incorporated into the resonator.

Origin of The Cw Chemical Hf/Df Laser

The possibility of the creation of infrared lasers based on the vibrationally excited products of a chemical reaction was first proposed by John Polanyi in 1961. A pulsed (although not chemical) laser was demonstrated by Jerome V. V. Kasper and George C. Pimentel in 1965. First, hydrogen chloride (HCl) was pumped optically so vigorously that the molecule disassociated and then re-combined, leaving it in an excited state suitable for a laser. Then hydrogen fluoride (HF) and deuterium fluoride (DF) were demonstrated. Pimentel went on to explore a DF-CO_2 transfer laser. Although this work did not produce a purely chemical continuous wave laser, it paved the way by showing the viability of the chemical reaction as a pumping mechanism for a chemical laser.

The continuous wave (CW) chemical HF laser was first demonstrated in 1969, and patented in 1972, by D. J. Spencer, T. A. Jacobs, H. Mirels and R. W. F. Gross at The Aerospace Corporation in El Segundo, California. This device used the mixing of adjacent streams of H_2 and F, within an optical cavity, to create vibrationally-excited HF that lased. The atomic fluorine was provided by dissociation of SF_6 gas using a DC electrical discharge. Later work at US Army, US Air Force, and US Navy contractor organizations (e.g. TRW) used a chemical reaction to provide the atomic fluorine, a concept included in the patent disclosure of Spencer et al. The latter configuration obviated the need for electrical power and led to the development of high-power lasers for military applications.

The analysis of the HF laser performance is complicated due to the need to simultaneously consider the fluid dynamic mixing of adjacent supersonic streams, multiple non-equilibrium chemical reactions and the interaction of the gain medium with the optical cavity. The researchers at The Aerospace Corporation developed the first exact analytic (flame sheet) solution, the first numerical computer code solution and the first simplified model describing CW HF chemical laser performance.

Chemical lasers stimulated the use of wave-optics calculations for resonator analysis. This work was pioneered by E. A. Sziklas (Pratt & Whitney) and A. E. Siegman (Stanford University). Part I of their work dealt with Hermite-Gaussian Expansion and has received little use compared with Part II, which dealt with the Fast Fourier transform method, which is now a standard tool at United Technologies Corporation (SOQ), Lockheed Martin (LMWOC), SAIC (ACS), Boeing (OSSIM), tOSC, MZA (Wave Train), and OPCI. Most of these companies competed for contracts to build HF and DF lasers for DARPA, the US Air Force, the US Army, or the US Navy throughout the 1970s and 1980s. General Electric and Pratt & Whitney dropped out of the competition in the early 1980s leaving the field to Rocketdyne (now ironically part of Pratt & Whitney - although the laser organization remains today with Boeing) and TRW (now part of Northrop Grumman).

Comprehensive chemical laser models were developed at SAIC by R. C. Wade, at TRW by C.-C. Shih, by D. Bullock and M. E. Lainhart, and at Rocketdyne by D. A. Holmes and T. R. Waite. Of these, perhaps the most sophisticated was the CROQ code at TRW, outpacing the early work at Aerospace Corporation.

Performance

The early analytical models coupled with chemical rate studies led to the design of efficient experimental CW HF laser devices at United Aircraft, and The Aerospace Corporation. Power levels up to 10 kW were achieved. DF lasing was obtained by the substitution of D_2 for H_2. A group at United Aircraft Research Laboratories produced a re-circulating chemical laser, which did not rely on the continuous consumption of chemical reactants.

The TRW Systems Group in Redondo Beach, California, subsequently received US Air Force contracts to build higher power CW HF/DF lasers. Using a scaled-up version of an Aerospace Corporation design, TRW achieved 100 kW power levels. General Electric, Pratt & Whitney, & Rocketdyne built various chemical lasers on company funds in anticipation of receiving DoD contracts to build even larger lasers. Only Rocketdyne received contracts of sufficient value to continue competing with TRW. TRW produced the MIRA-CL device for the U.S. Navy that achieved megawatt power levels. The latter is believed to be the highest power continuous laser, of any type, developed to date (2007).

TRW also produced a cylindrical chemical laser (the Alpha laser) for DARPA, which had the advantage, at least on paper, of being scalable to even larger powers. However, by 1990, the interest in chemical lasers had shifted toward shorter wavelengths, and the chemical oxygen-iodine laser (COIL) gained the most interest, producing radiation at 1.315 µm. There is a further advantage that the COIL laser generally produces single wavelength radiation, which is very helpful for forming a very well focused beam. This type of COIL laser is used today in the ABL (Airborne Laser, the laser itself being built by Northrop Grumman) and in the ATL (Advanced Tactical Laser) produced by Boeing. Meanwhile, a lower power HF laser was used for the THEL (Tactical High Energy Laser) built in the late 1990s for the Israeli Ministry of Defense in cooperation with the U.S. Army SMDC. It holds the distinction of being the first fielded high energy laser to demonstrate effectiveness in fairly realistic tests against rockets and artillery. The MIRACL laser has demonstrated effectiveness against certain targets flown in front of it at White Sands Missile Range, but it is not configured for actual service as a fielded weapon. ABL was successful in shooting down several full sized missiles from significant ranges, and ATL was successful in disabling moving land vehicles and other tactical targets.

Despite the performance advantages of chemical lasers, the Department of Defense stopped all development of chemical laser systems with the termination of the Airborne Laser Testbed in 2012. The desire for a "renewable" power source, i.e. not having to supply unusual chemicals like fluorine, deuterium, basic hydrogen-peroxide, or iodine, led the DoD to push for electrically pumped lasers such as diode pumped alkali lasers (DPALS).

Dye Laser

A dye laser is a laser which uses an organic dye as the lasing medium, usually as a liquid solution. Compared to gases and most solid state lasing media, a dye can usually be used for a much wider range of wavelengths, often spanning 50 to 100 nanometers or more. The wide bandwidth makes them particularly suitable for tunable lasers and pulsed lasers. The dye rhodamine 6G, for example, can be tuned from 635 nm (orangish-red) to 560 nm (greenish-yellow), and produce pulses as short as 16 femtoseconds. Moreover, the dye can be replaced by another type in order to generate an even broader range of wavelengths with the same laser, from the near-infrared to the near-ultraviolet, although this usually requires replacing other optical components in the laser as well.

Close-up of a table-top CW dye laser based on Rhodamine 6G, emitting at 580 nm (yellow). The emitted laser beam is visible as faint yellow lines between the yellow window (center) and the yellow optics (upper-right). The orange dye-solution enters the laser from the left and exits to the right, and is pumped by a 514 nm (blue-green) beam from an argon laser. The pump laser can be seen entering the dye jet, beneath the yellow window.

Dye lasers were independently discovered by P. P. Sorokin and F. P. Schäfer (and colleagues) in 1966.

In addition to the usual liquid state, dye lasers are also available as solid state dye lasers (SSDL). SSDL use dye-doped organic matrices as gain medium.

Construction

A dielectric mirror used in a dye laser.

A dye laser consists of an organic dye mixed with a solvent, which may be circulated through a dye cell, or streamed through open air using a dye jet. A high energy source of light is needed to 'pump' the liquid beyond its lasing threshold. A fast discharge flashlamp or an external laser is usually used for this purpose. Mirrors are also needed to oscillate the light produced by the dye's fluorescence, which is amplified with each pass through the liquid. The output mirror is normally around 80% reflective, while all other mirrors are usually more than 99.9% reflective. The dye solution is usually circulated at high speeds, to help avoid triplet absorption and to decrease degradation of the dye. A prism or diffraction grating is usually mounted in the beam path, to allow tuning of the beam.

Because the liquid medium of a dye laser can fit any shape, there are a multitude of different configurations that can be used. A Fabry–Pérot laser cavity is usually used for flashlamp pumped lasers, which consists of two mirrors, which may be flat or curved, mounted parallel to each other with the laser medium in between. The dye cell is usually side-pumped, with one or more flashlamps running parallel to the dye cell in a reflector cavity. The reflector cavity is often water cooled, to prevent thermal shock in the dye caused by the large amounts of near-infrared radiation which the flashlamp produces. Axial pumped lasers have a hollow, annular-shaped flashlamp that surrounds the dye cell, which has lower inductance for a shorter flash, and improved transfer efficiency. Coaxial pumped lasers have an annular dye cell that surrounds the flash lamp, for even better transfer efficiency, but have a lower gain due to diffraction losses. Flash pumped lasers can be used only for pulsed output applications.

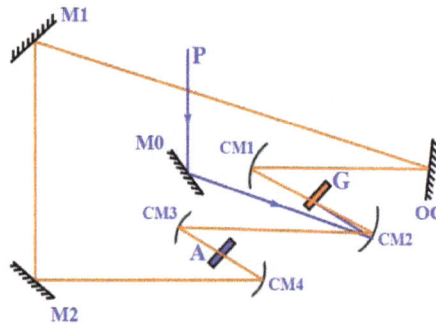

A ring dye laser. P-pump laser beam; G-gain dye jet; A-saturable absorber dye jet; M0, M1, M2-planar mirrors; OC–output coupler; CM1 to CM4-curved mirrors.

A ring laser design is often chosen for continuous operation, although a Fabry–Pérot design is sometimes used. In a ring laser, the mirrors of the laser are positioned to allow the beam to travel in a circular path. The dye cell, or cuvette, is usually very small. Sometimes a dye jet is used to help avoid reflection losses. The dye is usually pumped with an external laser, such as a nitrogen, excimer, or frequency doubled Nd:YAG laser. The liquid is circulated at very high speeds, to prevent triplet absorption from cutting off the beam. Unlike Fabry–Pérot cavities, a ring laser does not generate standing waves which cause spatial hole burning, a phenomenon where energy becomes trapped in unused portions of the medium between the crests of the wave. This leads to a better gain from the lasing medium.

Operation

The dyes used in these lasers contain rather large, organic molecules which fluoresce. The incoming light excites the dye molecules into the state of being ready to emit stimulated radiation; the singlet state. In this state, the molecules emit light via fluorescence, and the dye is transparent to the lasing wavelength. Within a microsecond or less, the molecules will change to their triplet state. In the triplet state, light is emitted via phosphorescence, and the molecules absorb the lasing wavelength, making the dye opaque. Liquid dyes also have an extremely high lasing-threshold. Flashlamp-pumped lasers need a flash with an extremely short duration, to deliver the large amounts of energy necessary to bring the dye past threshold before triplet absorption overcomes singlet emission. Dye lasers with an external pump-laser can direct enough energy of the proper wavelength into the dye with a relatively small amount of input energy, but the dye must be circulated at high speeds to keep the triplet molecules out of the beam path.

A cuvette used in a dye laser

Since organic dyes tend to decompose under the influence of light, the dye solution is normally circulated from a large reservoir. The dye solution can be flowing through a cuvette, i.e., a glass container, or be as a *dye jet*, i.e., as a sheet-like stream in open air from a specially-shaped nozzle. With a dye jet, one avoids reflection losses from the glass surfaces and contamination of the walls of the cuvette. These advantages come at the cost of a more-complicated alignment.

Liquid dyes have very high gain as laser media. The beam needs to make only a few passes through the liquid to reach full design power, and hence, the high transmittance of the output coupler. The high gain also leads to high losses, because reflections from the dye-cell walls or flashlamp reflector cause parasitic oscillations, dramatically reducing the amount of energy available to the beam. Pump cavities are often coated, anodized, or otherwise made of a material that will not reflect at the lasing wavelength while reflecting at the pump wavelength.

CW Dye Lasers

Continuous-wave (CW) dye lasers often use a dye jet. CW dye-lasers can have a linear or a ring cavity, and provided the foundation for the development of femtosecond lasers.

Narrow Linewidth Dye Lasers

Multiple prisms are often used to tune the output of a dye laser.

Dye lasers' emission is inherently broad. However, tunable narrow linewidth emission has been central to the success of the dye laser. In order to produce narrow bandwidth tuning these lasers use many types of cavities and resonators which include gratings, prisms, multiple-prism grating arrangements, and etalons.

The first narrow linewidth dye laser, introduced by Hänsch, used a Galilean telescope as beam expander to illuminate the diffraction grating. Next were the grazing-incidence grating designs and the multiple-prism grating configurations. The various resonators and oscillator designs developed for dye lasers have been successfully adapted to other laser types such as the diode laser. The physics of narrow-linewidth multiple-prism grating lasers was explained by Duarte and Piper.

Chemicals Used

Some of the laser dyes are rhodamine (orange, 540–680 nm), fluorescein (green, 530–560 nm), coumarin (blue 490–620 nm), stilbene (violet 410–480 nm), umbelliferone (blue, 450–470 nm), tetracene, malachite green, and others. While some dyes are actually used in food coloring, most dyes are very toxic, and often carcinogenic. Many dyes, such as rhodamine 6G, (in its chloride form), can be very corrosive to all metals except stainless steel. Although dyes have very broad fluorescence spectra, the dye's absorption and emission will tend to center on a certain wavelength and taper off to each side, forming a tunability curve, with the absorption center being of a shorter wavelength than the emission center. Rhodamine 6G, for example, has its highest output around 590 nm, and the conversion efficiency lowers as the laser is tuned to either side of this wavelength.

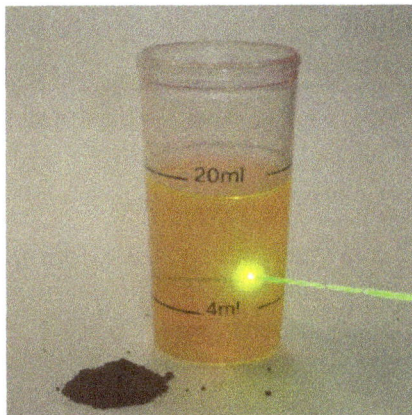

Rhodamine 6G Chloride powder; mixed with methanol; emitting yellow light under the influence of a green laser

A wide variety of solvents can be used, although some dyes will dissolve better in some solvents than in others. Some of the solvents used are water, glycol, ethanol, methanol, hexane, cyclohexane, cyclodextrin, and many others. Solvents are often highly toxic, and can sometimes be absorbed directly through the skin, or through inhaled vapors. Many solvents are also extremely flammable. The various solvents can also have an effect on the specific color of the dye solution and, thus, on the lasing bandwidth obtainable with a particular laser-pumping source.

Adamantane is added to some dyes to prolong their life.

Cycloheptatriene and cyclooctatetraene (COT) can be added as triplet quenchers for rhodamine G, increasing the laser output power. Output power of 1.4 kilowatt at 585 nm was achieved using Rhodamine 6G with COT in methanol-water solution.

Excitation Lasers

Flashlamps and several types of lasers can be used to optically pump dye lasers. A partial list of excitation lasers include:

- Copper vapor lasers
- Diode lasers
- Excimer lasers
- Nd:YAG lasers (mainly second and third harmonics)
- Nitrogen lasers
- Ruby lasers
- Argon ion lasers in the CW regime
- Krypton ion lasers in the CW regime

Ultra-Short Optical Pulses

R. L. Fork, B. I. Greene, and C. V. Shank demonstrated, in 1981, the generation of ultra-short laser pulse using a ring-dye laser (or dye laser exploiting colliding pulse mode-locking). Such kind of laser is capable of generating laser pulses of ~ 0.1 ps duration.

The introduction of grating techniques and intra-cavity prismatic pulse compressors eventually resulted in the routine emission of femtosecond dye laser pulses.

Applications

An atomic vapor laser isotope separation experiment at LLNL. Green light is from a copper vapor pump laser used to pump a highly tuned dye laser which is producing the orange light.

Dye lasers are very versatile. In addition to their recognized wavelength agility these lasers can offer very large pulsed energies or very high average powers. Flash-lamp-pumped dye lasers have been shown to yield hundreds of Joules per pulse and copper-laser-pumped dye lasers are known to yield average powers in the kilowatt regime.

Dye lasers are used in many applications including:

- astronomy (as laser guide stars),
- atomic vapor laser isotope separation
- manufacturing
- medicine
- spectroscopy

In laser medicine these lasers are applied in several areas, including dermatology where they are used to make skin tone more even. The wide range of wavelengths possible allows very close matching to the absorption lines of certain tissues, such as melanin or hemoglobin, while the narrow bandwidth obtainable helps reduce the possibility of damage to the surrounding tissue. They are used to treat port-wine stains and other blood vessel disorders, scars and kidney stones. They can be matched to a variety of inks for tattoo removal, as well as a number of other applications.

In spectroscopy, dye lasers can be used to study the absorption and emission spectra of various materials. Their tunability, (from the near-infrared to the near-ultraviolet), narrow bandwidth, and high intensity allows a much greater diversity than other light sources. The variety of pulse widths, from ultra-short, femtosecond pulses to continuous-wave operation, makes them suitable for a wide range of applications, from the study of fluorescent lifetimes and semiconductor properties to lunar laser ranging experiments.

Tunable lasers are used in swept-frequency metrology to enable measurement of absolute distances with very high accuracy. A two axis interferometer is set up and by sweeping the frequency, the frequency of the light returning from the fixed arm is slightly different from the frequency returning from the distance measuring arm. This produces a beat frequency which can be detected and used to determine the absolute difference between the lengths of the two arms.

Solid-State Laser

A solid-state laser is a laser that uses a gain medium that is a solid, rather than a liquid such as in dye lasers or a gas as in gas lasers. Semiconductor-based lasers are also in the solid state, but are generally considered as a separate class from solid-state lasers.

Solid-State Media

Generally, the active medium of a solid-state laser consists of a glass or crystalline "host" material to which is added a "dopant" such as neodymium, chromium, erbium, or ytterbium. Many of the common dopants are rare earth elements, because the excited states of such ions are not strongly coupled with the thermal vibrations of their crystal lattices (phonons), and their operational thresholds can be reached at relatively low intensities of laser pumping.

There are many hundreds of solid-state media in which laser action has been achieved, but relatively few types are in widespread use. Of these, probably the most common is neodymium-doped yttrium aluminum garnet (Nd:YAG). Neodymium-doped glass (Nd:glass) and ytterbium-doped glasses or ceramics are used at very high power levels (terawatts) and high energies (megajoules), for multiple-beam inertial confinement fusion.

The first material used for lasers was synthetic ruby crystals. Ruby lasers are still used for a few applications, but they are not common because of their low power efficiencies.

At room temperature, ruby lasers emit only short pulses of light, but at cryogenic temperatures they can be made to emit a continuous train of pulses.

Some solid-state lasers can also be tunable using several intracavity techniques which employ etalons, prisms, and gratings, or a combination of these. Titanium-doped sapphire is widely used for its broad tuning range, 660 to 1080 nanometers. Alexandrite lasers are tunable from 700 to 820 nm, and they yield higher-energy pulses than titanium-sapphire lasers because of the gain medium's longer energy storage time and higher damage threshold.

Pumping

Solid state lasing media are typically optically pumped, using either a flashlamp or arc lamp, or by laser diodes. Diode-pumped solid-state lasers tend to be much more efficient, and have become much more common as the cost of high power semiconductor lasers has decreased.

Mode Locking

Mode locking of solid state lasers has wide applications as large energy ultra-short pulses can be obtained. Like its counterpart, the fiber laser, there are three types of real saturable absorbers widely used as mode lockers: SESAM, SWCNT and graphene.

Particularly, graphene is a one-atom-thick planar sheet of sp2-bonded carbon atoms that are densely packed in a honeycomb crystal lattice. It is recently confirmed that the optical absorption from graphene could become saturated when the input optical intensity is above a threshold value. This nonlinear optical behavior is termed saturable absorption and the threshold value is called the saturation fluence. Graphene can be saturated readily under strong excitation over the visible to near-infrared region, due to the universal optical absorption and zero band gap. This has relevance for the mode locking of fiber lasers, where wideband tuneability may be obtained using graphene as the saturable absorber. Due to this special property, graphene has wide application in ultrafast photonics. Further, comparing with the SWCNTs, as graphene has a 2D structure it should have much smaller non-saturable loss and much higher damage threshold. Indeed, with an erbium-doped fiber laser, self-started mode locking and stable soliton pulse emission with high energy have been achieved.

History and Applications

Robert N. Hall developed the semiconductor laser in 1962, using GaAs (gallium arsenide), while working at General Electric in Schenectady, New York.

Solid-state lasers are being developed as optional weapons for the F-35 Lightning II, and are reaching near-operational status, as well as the introduction of Northrop Grumman's FIRESTRIKE laser weapon system. In April 2011 the United States Navy tested a high energy solid state laser. The exact range is classified, but they said it fired "miles not yards".

Uranium-doped calcium fluoride was the second type of solid state laser invented, in the 1960s. Peter Sorokin and Mirek Stevenson at IBM's laboratories in Yorktown Heights (US) achieved lasing at 2.5 μm shortly after Maiman's ruby laser.

Free-Electron Laser

The free-electron laser *FELIX* at the FOM Institute for Plasma Physics Rijnhuizen (nl), Nijmegen, The Netherlands.

A free-electron laser (FEL), is a type of laser whose lasing medium consists of very-high-speed electrons moving freely through a magnetic structure, hence the term *free electron*. The free-electron laser is tunable and has the widest frequency range of any laser type, currently ranging in wavelength from microwaves, through terahertz radiation and infrared, to the visible spectrum, ultraviolet, and X-ray.

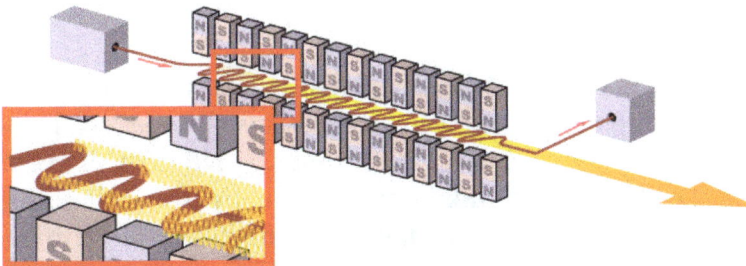

Schematic representation of an undulator, at the core of a free-electron laser.

The free-electron laser was invented by John Madey in 1971 at Stanford University. The free-electron laser utilizes technology developed by Hans Motz and his coworkers, who built an undulator at Stanford in 1953, using the wiggler magnetic configuration which is one component of a free electron laser. Madey used a 43-MeV electron beam and 5 m long wiggler to amplify a signal.

Beam Creation

To create a FEL, a beam of electrons is accelerated to almost the speed of light. The beam passes through an undulator, a side to side magnetic field produced by a periodic arrangement of magnets with alternating poles across the beam path. The direction of the beam is called the longitudinal direction, while the direction across the beam path is called transverse. This array of magnets is called an undulator or a wiggler, because it forces the electrons in the beam to wiggle transversely along a sinusoidal path about the axis of the undulator.

The undulator of *FELIX*.

The transverse acceleration of the electrons across this path results in the release of photons (synchrotron radiation), which are monochromatic but still incoherent, because the electromagnetic waves from randomly distributed electrons interfere constructively and destructively in time, and the resulting radiation power scales linearly with the number of electrons. If an external laser is provided or if the synchrotron radiation becomes sufficiently strong, the transverse electric field of the radiation beam interacts with the transverse electron current created by the sinusoidal wiggling motion, causing some electrons to gain and others to lose energy to the optical field via the ponderomotive force.

This energy modulation evolves into electron density (current) modulations with a period of one optical wavelength. The electrons are thus clumped, called *microbunches*, separated by one optical wavelength along the axis. Whereas conventional undulators would cause the electrons to radiate independently, the radiation emitted by the bunched electrons are in phase, and the fields add together coherently.

The FEL radiation intensity grows, causing additional microbunching of the electrons, which continue to radiate in phase with each other. This process continues until the electrons are completely microbunched and the radiation reaches a saturated power several orders of magnitude higher than that of the undulator radiation.

The wavelength of the radiation emitted can be readily tuned by adjusting the energy of the electron beam or the magnetic-field strength of the undulators.

FELs are relativistic machines. The wavelength of the emitted radiation, λ_r, is given by

$$\lambda_r = \frac{\lambda_u}{2\gamma^2}(1+K^2),$$

or when the wiggler strength parameter K, discussed below, is small

$$\lambda_r \propto \frac{\lambda_u}{2\gamma^2},$$

where λ_u is the undulator wavelength (the spatial period of the magnetic field), γ is the relativistic Lorentz factor and the proportionality constant depends on the undulator geometry and is of the order of 1.

This formula can be understood as a combination of two relativistic effects. Imagine you are sitting on an electron passing through the undulator. Due to Lorentz contraction the undulator is shortened by a γ factor and the electron experiences much shorter undulator wavelength λ_u/γ. However, the radiation emitted at this wavelength is observed in the laboratory frame of reference and the relativistic Doppler effect brings the second γ factor to the above formula. Rigorous derivation from Maxwell's equations gives the divisor of 2 and the proportionality constant. In an X-ray FEL the typical undulator wavelength of 1 cm is transformed to X-ray wavelengths on the order of 1 nm by $\gamma \approx 2000$, i.e. the electrons have to travel with the speed of 0.9999998c.

Wiggler Strength Parameter K

K, a dimensionless parameter, tells the wiggler strength as the relationship between the length of a period and the radius of bend,

$$K = \frac{\gamma\lambda_u}{\tau\rho} = \frac{eB_0\lambda_u}{\sqrt{4\tau}m_e c}$$

where ρ is the bending radius, B_0 is the applied magnetic field and m m_e the electron mass.

Quantum Effects

In most cases, the theory of classical electromagnetism adequately accounts for the behavior of free electron lasers. For sufficiently short wavelengths, quantum effects of electron recoil and shot noise may have to be considered.

Large Facilities Required

Free-electron lasers require the use of an electron accelerator with its associated shielding, as accelerated electrons can be a radiation hazard if not properly contained. These accelerators are typically powered by klystrons, which require a high voltage supply. The electron beam must be maintained in a vacuum which requires the use of numerous vacuum pumps along the beam path. While this equipment is bulky and expensive, free-electron lasers can achieve very high peak powers, and the tunability of FELs makes them highly desirable in many disciplines, including chemistry, structure determination of molecules in biology, medical diagnosis, and nondestructive testing.

X-ray Laser Without Mirrors

The lack of a material to make mirrors that can reflect extreme ultraviolet and x-rays means that FELs at these frequencies cannot use a resonant cavity like other lasers, which reflects the radiation so it makes multiple passes through the undulator. Consequently, in an X-ray FEL the output beam is produced by a single pass of radiation through the undulator; there must be enough amplification over a single pass to produce an adequately bright beam.

X-ray free electron lasers use long undulators. The underlying principle of the intense pulses from the X-ray laser lies in the principle of self-amplified spontaneous emission (SASE), which leads to the microbunching. Initially all electrons are distributed evenly and they emit incoherent spontaneous radiation only. Through the interaction of this radiation and the electrons' oscillations, they drift into microbunches separated by a distance equal to one radiation wavelength. Through this interaction, all electrons begin emitting coherent radiation in phase. All emitted radiation can reinforce itself perfectly whereby wave crests and wave troughs are always superimposed on one another in the best possible way. This results in an exponential increase of emitted radiation power, leading to high beam intensities and laser-like properties. Examples of facilities operating on the SASE FEL principle include the Free electron LASer in Hamburg (FLASH), the Linac Coherent Light Source (LCLS) at the SLAC National Accelerator Laboratory, the European x-ray free electron laser (XFEL) in Hamburg, the SPring-8 Compact SASE Source (SCSS), the SwissFEL at the Paul Scherrer Institute (Switzerland) and, as of 2011, the SACLA at the RIKEN Harima Institute in Japan.

Self Seeding

One problem with SASE FELs is the lack of temporal coherence due to a noisy start-up process. To avoid this, one can "seed" an FEL with a laser tuned to the resonance of the FEL. Such a temporally coherent seed can be produced by more conventional means, such as by high-harmonic generation (HHG) using an optical laser pulse. This results in coherent amplification of the input signal; in effect, the output laser quality is characterized by the seed. While HHG seeds are available at wavelengths down to the extreme ultraviolet, seeding is not feasible at x-ray wavelengths due to the lack of conventional x-ray lasers. In late 2010, in Italy, the seeded-FEL source FERMI@ Elettra started commissioning, at the Sincrotrone Trieste Laboratory. FERMI@Elettra is a single-pass FEL user-facility covering the wavelength range from 100 nm (12 eV) to 10 nm (124 eV), located next to the third-generation synchrotron radiation facility ELETTRA in Trieste, Italy. The advent of femtosecond lasers has revolutionized many areas of science from solid state physics to biology.

In 2012, scientists working on the LCLS overcame the seeding limitation for x-ray wavelengths by self-seeding the laser with its own beam after being filtered through a diamond monochromator. The resulting intensity and monochromaticity of the beam were unprecedented and allowed new experiments to be conducted involving manipulating atoms and imaging molecules. Other labs around the world are incorporating the technique into their equipment.

Applications

Medical

Surgery

Research by Glenn Edwards and colleagues at Vanderbilt University's FEL Center in 1994 found that soft tissues including skin, cornea, and brain tissue could be cut, or ablated, using infrared FEL wavelengths around 6.45 micrometres with minimal collateral damage to adjacent tissue. This led to surgeries on humans, the first ever using a free-electron laser. Starting in 1999, Copeland and Konrad performed three surgeries in which they resected meningioma brain tumors. Beginning in 2000, Joos and Mawn performed five surgeries that cut a window in the sheath of the optic nerve, to test the efficacy for optic nerve sheath fenestration. These eight surgeries produced results consistent with the standard of care and with the added benefit of minimal collateral damage. A review of FELs for medical uses is given in the 1st edition of Tunable Laser Applications.

Fat Removal

Several small, clinical lasers tunable in the 6 to 7 micrometre range with pulse structure and energy to give minimal collateral damage in soft tissue were created. At Vanderbilt, there exists a Raman shifted system pumped by an Alexandrite laser.

Rox Anderson proposed the medical application of the free-electron laser in melting fats without harming the overlying skin. At infrared wavelengths, water in tissue was heated by the laser, but at wavelengths corresponding to 915, 1210 and 1720 nm, subsurface lipids were differentially heated more strongly than water. The possible applications of this selective photothermolysis (heating tissues using light) include the selective destruction of sebum lipids to treat acne, as well as targeting other lipids associated with cellulite and body fat as well as fatty plaques that form in arteries which can help treat atherosclerosis and heart disease.

Biology

Exceptionally bright and fast X-rays can image proteins using a sheet just one molecule thick. This technique allows first-time imaging of proteins that do not stack in a way that allows imaging by conventional techniques, 25% of the total number of proteins. Resolutions of 0.8 nm have been achieved with pulse durations of 30 femtoseconds. To get a clear view, a resolution of 0.1–0.3 nm is required. The short pulse durations prevented the lasers from destroying the molecules. The bright, fast X-rays were produced at the Linac Coherent Light Source at SLAC. As of 2014 LCLS was the world's most powerful X-ray FEL.

Military

FEL technology is being evaluated by the US Navy as a candidate for an antiaircraft and anti-missile directed-energy weapon. The Thomas Jefferson National Accelerator Facility's FEL has demonstrated over 14 kW power output. Compact multi-megawatt class FEL weapons are undergoing research. On June 9, 2009 the Office of Naval Research announced it had awarded Raytheon a contract to develop a 100 kW experimental FEL. On March 18, 2010 Boeing Directed Energy Systems announced the completion of an initial design for U.S. Naval use. A prototype FEL system was demonstrated, with a full-power prototype scheduled by 2018.

Gas Dynamic Laser

A gas dynamic laser (GDL) is a laser based on differences in relaxation velocities of molecular vibrational states. The lasing medium gas has such properties that an energetically lower vibrational state relaxes faster than a higher vibrational state, and so a population inversion is achieved in a particular time. It was invented by Edward Gerry and Arthur Kantrowitz at Avco Everett Research Laboratory in 1966.

Pure gas dynamic lasers usually use a combustion chamber, supersonic expansion nozzle, and CO_2, in a mixture with nitrogen or helium, as the laser medium.

Gas dynamic lasers can be pumped by combustion or adiabatic expansion of gas. Any hot and compressed gas with appropriate vibrational structure could be utilized.

The explosively pumped gas dynamic laser is a version of GDL pumped by expansion of explosion products. Hexanitrobenzene and/or tetranitromethane with metal powder is the preferred explosive. This device could have very high pulsed peak power output suitable for laser weapons.

Function

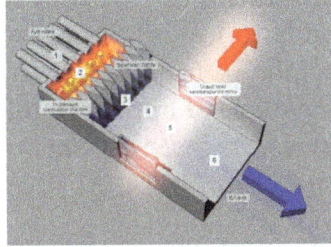

Gas dynamic laser components and function

- Hot compressed gas is generated.

- Gas expands through subsonic or supersonic expansion nozzle, the temperature of the gas becomes lower and according to Maxwell–Boltzmann distribution the gas isn't in thermodynamic equilibrium until the vibrational states relax.

- The gas flows through the tube of a particular length for a particular time. In this time lower vibrational state does relax but higher vibrational state doesn't. Thus population inversion is achieved.

- Gas flows through mirror area where stimulated emission takes place.

- Gas returns to equilibrium and becomes warm. It must be removed from the laser cavity or it will interfere with the thermodynamics and vibrational state relaxation of the freshly expanded gas.

Application

Almost any chemical laser uses gas-dynamic processes to increase its efficiency.

High energy efficiency (as high as 30%) and very high power output make GDL suitable for some (especially military) applications.

Raman Laser

A Raman laser is a specific type of laser in which the fundamental light-amplification mechanism is stimulated Raman scattering. In contrast, most "conventional" lasers (such as the ruby laser) rely on stimulated electronic transitions to amplify light.

Specific Properties of Raman Lasers

Spectral Flexibility

Raman lasers are optically pumped. However, this pumping does not produce a population inversion as in conventional lasers. Rather, pump photons are absorbed and "immediately" re-emitted as lower-frequency laser-light photons ("Stokes" photons) by stimulated Raman scattering. The difference between the two photon energies is fixed and corresponds to a vibrational frequency of the gain medium. This makes it possible, in principle, to produce arbitrary laser-output wavelengths by choosing the pump-laser wavelength appropriately. This is in contrast to conventional lasers, in which the possible laser output wavelengths are determined by the emission lines of the gain material.

In optical fibers made of silica, for example, the frequency shift corresponding to the largest Raman gain is about 13.2 THz. In the near infrared, this corresponds to a wavelength separation between pump light and laser-output light of about 100 nm.

Types of Raman Lasers

The first Raman laser, realized in 1962, used nitrobenzene as the gain medium, which was intra-cavity-pumped inside a Q-switching ruby laser. Various other gain media can be used to construct Raman lasers:

Raman Fiber Lasers

The first continuous-wave Raman laser using an optical fiber as the gain medium has been demonstrated in 1976. In fiber-based lasers, tight spatial confinement of the pump light is maintained over relatively large distances. This significantly lowers threshold pump powers down to practical levels and furthermore enables continuous-wave operation.

In 1988, the first Raman fiber laser based on fiber Bragg gratings has been made. Fiber Bragg gratings are narrow-band reflectors and act as the mirrors of the laser cavity. They are inscribed directly into the core of the optical fiber used as the gain medium, which eliminates substantial losses that previously arose due to the coupling of the fiber to external bulk-optic cavity reflectors.

Nowadays, commercially available fiber-based Raman lasers can deliver output powers in the range of a few tens of Watts in continuous-wave operation. A technique that is commonly employed in these devices is *cascading*, first proposed in 1994: The "first-order" laser light that is generated from the pump light in a single frequency-shifting step remains trapped in the laser resonator and is pushed to such high power levels that it acts itself as the pump for the generation of "second-order" laser light that is shifted by the same vibrational frequency again. In this way, a single laser resonator is used to convert the pump light (typically around 1060 nm) through several discrete steps to an "arbitrary" desired output wavelength.

Silicon Raman Lasers

More recently, Raman lasing has been demonstrated in silicon-based integrated-optical waveguides by Bahram Jalali's group at the University of California in Los Angeles in 2004 (pulsed operation) and by Intel in 2005 (continuous-wave), respectively. These developments received much attention because it was the first time that a laser was realized in silicon: "classical" lasing based on electronic transitions is prohibited in crystalline silicon due to its indirect bandgap. Practical silicon-based light sources would be very interesting for the field of silicon photonics, which seeks to exploit silicon not only for realizing electronics but also for novel light-processing functionality on the same chip.

Nuclear Pumped Laser

A nuclear pumped laser is a laser pumped with the energy of fission fragments. The lasing medium is enclosed in a tube lined with uranium-235 and subjected to high neutron flux in a nuclear reactor core. The fission fragments of the uranium create excited plasma with inverse population of energy levels, which then lases. Other methods, e.g. the He-Ar laser, can use the He(n,p)H reaction, the transmutation of helium-3 in a neutron flux, as the energy source, or employing the energy of the alpha particles.

This technology may achieve high excitation rates with small laser volumes.

Some example lasing media:

- carbon dioxide
- helium-argon
- helium-krypton
- helium-xenon

Development

Research in nuclear pumped lasers started in the early 1970s when researchers were unable to produce a laser with a wavelength shorter than 110 nm with the end goal of creating an x-ray laser. When laser wavelengths become that short the laser requires a huge amount of energy which must also be delivered in an extremely short period of time. In 1975 it was estimated, by George Chapline and Lowell Wood from the Lawrence Livermore National Laboratory, that "pumping a 10-keV (0.12-nm) laser would require around a watt per atom" in a pulse that was "10^{-15} seconds x the square of the wavelength in angstroms." As this problem was unsolvable with the materials at hand and a laser oscillator was not working, research moved to creating pumps that used excited plasma. Early attempts used high-powered lasers to excite the plasma to create an even more highly powered laser. Results using this method were unsatisfying, and fell short of the goal. Livermore scientists first suggested using a nuclear reaction as a power source in 1975. By 1980 Livermore considered both nuclear bombs and nuclear reactors as viable energy sources for an x-ray laser.

On November 14, 1980, the first successful test of the bomb-powered x-ray laser was conducted. The use of a bomb was initially supported over that of the reactor driven laser because it delivered a more intense beam. Livermore's research was almost entirely devoted to missile defense using x-ray lasers. The idea was to mount a system of nuclear bombs in space where these bombs would each power approximately 50 lasers. Upon detonation these lasers would fire and theoretically destroy several dozen incoming nuclear missiles at once. Opponents of this plan found many faults in such an approach and questioned aspects such as the power, range, accuracy, politics, and cost of such deployments. In 1985 a test titled 'Goldstone' revealed the delivered power to be less than believed. Efforts to focus the laser also failed.

Fusion lasers (reactor driven lasers) started testing after the bomb-driven lasers proved successful. While prohibitively expensive (estimated at 30,000 dollars per test), research was easier in that tests could be performed several times a day and the equipment could be reused. In 1984, a test achieved wavelengths of less than 21 nm the closest to an official x-ray laser yet. (There are many definitions for an x-ray laser, some of which require a wavelength of less than 10 nm). The Livermore method was to remove the outer electrons in heavy atoms to create a "neon-like" substance. When presented at an American Physical Society meeting, the success of the test was shared by an experiment from Princeton University which was better in size, cost, measured wavelength, and amplification than Livermore's test. Research has continued in the field of nuclear pumped lasers and it remains on the cutting edge of the field.

Uses

At least 3 uses for bomb pumped lasers have been proposed.

Propulsion

Laser propulsion is an alternative method of propulsion ideal for launching objects into orbit, as this method requires less fuel, meaning less mass must be launched. A nuclear pumped laser is ideal for this operation. A launch using laser propulsion requires high intensity, short pulses, good quality, and a high power output. A nuclear pumped laser would theoretically be capable of meeting these requirements.

Manufacturing

The characteristics of the nuclear pumped laser make it ideal for applications in deep-cut welding, cutting thick materials, the heat treating of metals, vapor deposition of ceramics, and the production of sub-micron sized particles.

Weapon

Titled Project Excalibur, the program was a part of President Reagan's Strategic Defense Initiative. Livermore Laboratories conceived of the initial idea and Edward Teller developed and presented the idea to the president. Permission was granted to pursue the project though it has been reported Reagan was reluctant to incorporate nuclear devices in the nation's plan against nuclear devices. While initial tests were promising, the results never reached acceptable levels. Later, lead scientists were accused of falsifying the reports. Project Excalibur was cancelled several years later.

Excimer Laser

An excimer laser, sometimes more correctly called an exciplex laser, is a form of ultraviolet laser which is commonly used in the production of microelectronic devices, semiconductor based integrated circuits or "chips", eye surgery, and micromachining.

Terminology

The term excimer is short for 'excited dimer', while exciplex is short for 'excited complex'. Most "excimer" lasers are of the noble gas halide type, for which the term *excimer* is strictly speaking a misnomer (since a dimer refers to a molecule of two identical or similar parts): The correct but less commonly used name for such is exciplex laser.

History

The excimer laser was invented in 1970 by Nikolai Basov, V. A. Danilychev and Yu. M. Popov, at the Lebedev Physical Institute in Moscow, using a xenon dimer (Xe_2) excited by an electron beam to give stimulated emission at 172 nm wavelength. A later improvement, developed by many groups in 1975 was the use of noble gas halides (originally Xe Br). These groups include the Avco Everett Research Laboratory, Sandia Laboratories, the Northrop Research and Technology Center, and the United States Government's Naval Research Laboratory who also developed a XeCl Laser that was excited using a microwave discharge.

Construction

An excimer laser typically uses a combination of a noble gas (argon, krypton, or xenon) and a reactive gas (fluorine or chlorine). Under the appropriate conditions of electrical stimulation and high pressure, a pseudo-molecule called an excimer (or in the case of noble gas halides, exciplex) is created, which can only exist in an energized state and can give rise to laser light in the ultraviolet range.

Operation

Laser action in an excimer molecule occurs not because it has a bound (associative) excited state, but a repulsive (dissociative) ground state. This is because noble gases such as xenon and krypton are highly inert and do not usually form chemical compounds. However, when in an excited state (induced by an electrical discharge or high-energy electron beams, which produce high energy pulses), they can form temporarily bound molecules with themselves (dimers) or with halogens (complexes) such as fluorine and chlorine. The excited compound can give up its excess energy by undergoing spontaneous or stimulated emission, resulting in a strongly repulsive ground state molecule which very quickly (on the order of a picosecond) dissociates back into two unbound atoms. This forms a population inversion.

Wavelength Determination

The wavelength of an excimer laser depends on the molecules used, and is usually in the ultraviolet:

Excimer	Wavelength	Relative power mW
Ar_2^*	126 nm	
Kr_2^*	146 nm	
F_2^*	157 nm	
Xe_2^*	172 & 175 nm	
ArF	193 nm	60
KrF	248 nm	100
XeBr	282 nm	
XeCl	308 nm	50
XeF	351 nm	45
KrCl	222 nm	25

Excimer lasers, such as XeF and KrF, can also be made slightly *tunable* using a variety of prism and grating intracavity arrangements.

Repetition Rate of Excimer Laser

Excimer lasers are usually operated with a pulse repetition rate of around 100 Hz and a pulse duration of ~10 ns, although some operate at pulse repetition rates as high as 8 kHz and some have pulsewidths as large as 30 ns.

Major Applications

The ultraviolet light from an excimer laser is well absorbed by biological matter and organic compounds. Rather than burning or cutting material, the excimer laser adds enough energy to disrupt the molecular bonds of the surface tissue, which effectively disintegrates into the air in a tightly controlled manner through ablation rather than burning. Thus excimer lasers have the useful property that they can remove exceptionally fine layers of surface material with almost no heating or change to the remainder of the material which is left intact. These properties make excimer lasers well suited to precision micromachining organic material (including certain polymers and plastics), or delicate surgeries such as eye surgery LASIK.

Photolithography

Excimer lasers are widely used in high-resolution photolithography machines, one of the critical technologies required for microelectronic chip manufacturing. Current state-of-the-art lithography tools use deep ultraviolet (DUV) light from the KrF and ArF excimer lasers with wavelengths of 248 and 193 nanometers (the dominant lithography technology today is thus also called "excimer laser lithography"), which has enabled transistor feature sizes to shrink below 45 nanometers. Excimer laser lithography has thus played a critical role in the continued advance of the so-called Moore's law for the last 20 years.

The most widespread industrial application of excimer lasers has been in deep-ultraviolet photolithography, a critical technology used in the manufacturing of microelectronic devices (i.e., semiconductor integrated circuits or "chips"). Historically, from the early 1960s through the mid-1980s, mercury-xenon lamps had been used in lithography for their spectral lines at 436, 405 and 365 nm wavelengths. However, with the semiconductor industry's need for both higher resolution (to produce denser and faster chips) and higher throughput (for lower costs), the lamp-based lithography tools were no longer able to meet the industry's requirements. This challenge was overcome when in a pioneering development in 1982, deep-UV excimer laser lithography was proposed and demonstrated at IBM by Kanti Jain. With phenomenal advances made in equipment technology in the last two decades, and today microelectronic devices fabricated using excimer laser lithography totaling $400 billion in annual production, it is the semiconductor industry view that excimer laser lithography has been a crucial factor in the continued advance of Moore's law, enabling minimum features sizes in chip manufacturing to shrink from 800 nanometers in 1990 to 15 nanometers in 2016. This trend is expected to continue into this decade for even denser chips, with minimum features approaching 10 nanometers. From an even broader scientific and technological perspective, since the invention of the laser in 1960, the development of excimer laser lithography has been highlighted as one of the major milestones in the 50-year history of the laser.

Medical Uses

The high-power ultraviolet output of excimer lasers also makes them useful for surgery (particularly eye surgery) and for dermatological treatment. In 1980–1983, Rangaswamy Srinivasan, Samuel Blum and James J. Wynne at IBM's T. J. Watson Research Center observed the effect of the ultraviolet excimer laser on biological materials. Intrigued, they investigated further, finding that the laser made clean, precise cuts that would be ideal for delicate surgeries. This resulted in a fundamental patent and Srinivasan, Blum and Wynne were elected to the National Inventors Hall of Fame in 2002. In 2012, the team members were honored with National Medal of Technology and Innovation by the President of The United States Barack Obama for their work related to the excimer laser. Subsequent work introduced the excimer laser for use in angioplasty. Xenon chloride (308 nm) excimer lasers can also treat a variety of dermatological conditions including psoriasis, vitiligo, atopic dermatitis, alopecia areata and leukoderma.

As light sources, excimer lasers are generally large in size, which is a disadvantage in their medical applications, although their sizes are rapidly decreasing with ongoing development.

Scientific Research

Excimer lasers are also widely used in numerous fields of scientific research, both as primary sources and, particularly the XeCl laser, as pump sources for tunable dye lasers, mainly to excite laser dyes emitting in the blue-green region of the spectrum.

Fiber Laser

A fiber laser or fibre laser is a laser in which the active gain medium is an optical fiber doped with rare-earth elements such as erbium, ytterbium, neodymium, dysprosium, praseodymium, thulium and holmium. They are related to doped fiber amplifiers, which provide light amplification without lasing. Fiber nonlinearities, such as stimulated Raman scattering or four-wave mixing can also provide gain and thus serve as gain media for a fiber laser.

Advantages and Applications

A laser cutting machine with a 2 kW continuous wave fiber laser

The advantages of fiber lasers over other types include:

- Light is already coupled into a flexible fiber: The fact that the light is already in a fiber allows it to be easily delivered to a movable focusing element. This is important for laser cutting, welding, and folding of metals and polymers.

- High output power: Fiber lasers can have active regions several kilometers long, and so can provide very high optical gain. They can support kilowatt levels of continuous output power because of the fiber's high surface area to volume ratio, which allows efficient cooling.

- High optical quality: The fiber's waveguiding properties reduce or eliminate thermal distortion of the optical path, typically producing a diffraction-limited, high-quality optical beam.

- Compact size: Fiber lasers are compact compared to rod or gas lasers of comparable power, because the fiber can be bent and coiled to save space.

- Reliability: Fiber lasers exhibit high temperature and vibrational stability, extended lifetime, and maintenance-free turnkey operation.

- High peak power and nanosecond pulses enable effective marking and engraving.

- The additional power and better beam quality provide cleaner cut edges and faster cutting speeds.

- Lower cost of ownership.

- Fiber lasers are now being used to make high-performance surface-acoustic wave (SAW) devices. These lasers raise throughput and lower cost of ownership in comparison to older solid-state laser technology.

Fiber laser can also refer to the machine tool that includes the fiber resonator.

Applications of fiber lasers include material processing (marking, engraving, cutting), telecommunications, spectroscopy, medicine, and directed energy weapons.

Design and Manufacture

Unlike most other types of lasers, the laser cavity in fiber lasers is constructed monolithically by fusion splicing different types of fiber; fiber Bragg gratings replace conventional dielectric mirrors to provide optical feedback. Another type is the single longitudinal mode operation of ultra narrow distributed feedback lasers (DFB) where a phase-shifted Bragg grating overlaps the gain medium. Fiber lasers are pumped by semiconductor laser diodes or by other fiber lasers. Q-switched pulsed fiber lasers offer a compact, electrically efficient alternative to Nd:YAG technology.

Double-Clad Fibers

Many high-power fiber lasers are based on double-clad fiber. The gain medium forms the core of the fiber, which is surrounded by two layers of cladding. The lasing mode propagates in the core, while a multimode pump beam propagates in the inner cladding layer. The outer cladding keeps this pump light confined. This arrangement allows the core to be pumped with a much higher-power beam than could otherwise be made to propagate in it, and allows the conversion of pump light with relatively low brightness into a much higher-brightness signal. As a result, fiber lasers and amplifiers are occasionally referred to as "brightness converters." There is an important question about the shape of the double-clad fiber; a fiber with circular symmetry seems to be the worst possible design. The design should allow the core to be small enough to support only a few (or even one) modes. It should provide sufficient cladding to confine the core and optical pump section over a relatively short piece of the fiber.

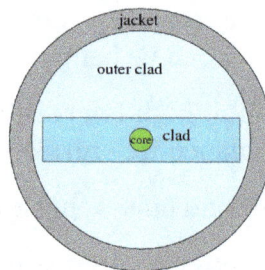

Double-clad fiber

Power Scaling

10,000W SM Laser

Recent developments in fiber laser technology have led to a rapid and large rise in achieved diffraction-limited beam powers from diode-pumped solid-state lasers. Due to the introduction of large mode area (LMA) fibers as well as continuing advances in high power and high brightness diodes, continuous-wave single-transverse-mode powers from Yb-doped fiber lasers have increased from 100 W in 2001 to >20 kW. Commercial single-mode lasers have reached 10 kW in CW power. In 2014 a combined beam fiber laser demonstrated power of 30 kW.

Mode Locking

Nonlinear Polarization Rotation

When linearly polarized light is incident to a piece of weakly birefringent fiber, the polarization of the light will generally become elliptically polarized in the fiber. The orientation and ellipticity of the final light polarization is fully determined by the fiber length and its birefringence. However, if the intensity of the light is strong, the non-linear optical Kerr effect in the fiber must be considered, which introduces extra changes to the light polarization. As the polarization change introduced by the optical Kerr effect depends on the light intensity, if a polarizer is put behind the fiber, the light intensity transmission through the polarizer will become light intensity dependent. Through appropriately selecting the orientation of the polarizer or the length of the fiber, an artificial saturable absorber effect with ultra-fast response could then be achieved in such a system, where light of higher intensity experiences less absorption loss on the polarizer. The NPR technique makes use of this artificial saturable absorption to achieve the passive mode locking in a fiber laser. Once a mode-locked pulse is formed, the non-linearity of the fiber further shapes the pulse into an optical soliton and consequently the ultrashort soliton operation is obtained in the laser. Soliton operation is almost a generic feature of the fiber lasers mode-locked by this technique and has been intensively investigated.

Semiconductor Saturable Absorber Mirrors (Sesams)

Semiconductor saturable absorbers were used for laser mode-locking as early as 1974 when p-type germanium is used to mode lock a CO_2 laser which generated pulses ~500 ps . Modern SESAMs are III-V semiconductor single quantum well (SQW) or multiple quantum wells grown on semiconductor distributed Bragg reflectors (DBRs). They were initially used in a Resonant Pulse Modelocking (RPM) scheme as starting mechanisms for Ti:Sapphire lasers which employed KLM as a fast saturable absorber . RPM is another coupled-cavity mode-locking technique. Different from APM lasers which employ non-resonant Kerr-type phase nonlinearity for pulse shortening, RPM employs the amplitude nonlinearity provided by the resonant band filling effects of semiconductors. SESAMs were soon developed into intracavity saturable absorber devices because of more inherent simplicity with this structure. Since then, the use of SESAMs has enabled the pulse durations, average powers, pulse energies and repetition rates of ultrafast solid-state lasers to be improved by several orders of magnitude. Average power of 60 W and repetition rate up to 160 GHz were obtained. By using SESAM-assisted KLM, sub-6 fs pulses directly from a Ti: Sapphire oscillator was achieved. A major advantage SESAMs have over other saturable absorber techniques is that absorber parameters can be easily controlled over a wide range of values. For example, saturation fluence can be controlled by varying the reflectivity of the top reflector while modulation depth and recovery time can be tailored by changing the low temperature growing conditions for the absorber layers . This freedom of design has further extended the application of SESAMs into modelocking of fiber lasers where a relatively high modulation depth is needed to ensure self-starting and operation stability. Fiber lasers working at ~ 1 μm and 1.5 μm were successfully demonstrated.

Carbon Nanotube Saturable Absorbers

Graphene Saturable Absorbers

Graphene is a one-atom-thick planar sheet of sp2-bonded carbon atoms that are densely packed in a honeycomb crystal lattice. Optical absorption from graphene can become saturated when the input optical intensity is above a threshold value. This nonlinear optical behavior is termed saturable absorption and the threshold value is called the saturation fluency. Graphene can be saturated readily under strong excitation over the visible to near-infrared region, due to the universal optical absorption and zero band gap. This has relevance for the mode locking of fiber lasers, where wideband tunability may be obtained using graphene as the saturable absorber. Due to this special property, graphene has wide application in ultrafast photonics. Furthermore, comparing with the SWCNTs, as graphene has a 2D structure it should have much smaller non-saturable loss and much higher damage threshold. Self-started mode locking and stable soliton pulse emission with high energy have been achieved with a graphene saturable absorber in an erbium-doped fiber laser. Atomic layer graphene possesses wavelength-insensitive ultrafast saturable absorption, which can be exploited as a "full-band" mode locker. With an erbium-doped dissipative soliton fiber laser mode locked with few layer graphene, it has been experimentally shown that dissipative solitons with continuous wavelength tuning as large as 30 nm (1570–1600 nm) can be obtained.

Active Mode Locking

Active mode-locking is normally achieved by modulating the loss (or gain) of the laser cavity at a repetition rate equivalent to the cavity frequency, or a harmonic thereof. In practice, the modulator can be acousto-optic or electro-optic modulator, Mach-Zehnder integrated-optic modulators, or a semiconductor electro-absorption modulator (EAM). The principle of active mode-locking with a sinusoidal modulation. In this situation, optical pulses will form in such a way as to minimize the loss from the modulator. The peak of the pulse would automatically adjust in phase to be at the point of minimum loss from the modulator. Because of the slow variation of sinusoidal modulation, it is not very straightforward for generating ultrashort optical pulses (< 1ps) using this method.

For stable operation, the cavity length must precisely match the period of the modulation signal or some integer multiple of it. The most powerful technique to solve this is regenerative mode locking i.e. a part of the output signal of the mode-locked laser is detected; the beatnote at the round-trip frequency is filtered out from the detector, and sent to an amplifier, which drives the loss modulator in the laser cavity. This procedure enforces synchronism if the cavity length undergoes fluctuations due to acoustic vibrations or thermal expansion. By using this method, highly stable mode-locked lasers have been achieved. The major advantage of active mode-locking is that it allows synchronized operation of the mode-locked laser to an external radio frequency (RF) source. This is very useful for optical fiber communication where synchronization is normally required between optical signal and electronic control signal. Also active mode-locked fiber can provide much higher repetition rate than passive mode-locking. Currently, fiber lasers and semiconductor diode lasers are the two most important types of lasers where active mode-locking are applied.

Dark Soliton Fiber Lasers

In the non-mode locking regime,the first dark soliton fiber laser has been successfully achieved in an all-normal dispersion erbium-doped fiber laser with a polarizer in cavity. Experimentally finding that apart from the bright pulse emission, under appropriate conditions the fiber laser could also emit single or multiple dark pulses. Based on numerical simulations we interpret the dark pulse formation in the laser as a result of dark soliton shaping.

Multiwavelength Fiber Lasers

Recently,multiwavelength dissipative soliton in an all normal dispersion fiber laser passively mode-locked with a SESAM has been generated. It is found that depending on the cavity birefringence, stable single-, dual- and triple-wavelength dissipative soliton can be formed in the laser. Its generation mechanism can be traced back to the nature of dissipative soliton.

Fiber Disk Lasers

Another type of fiber laser is the fiber disk laser. In such lasers, the pump is not confined within the cladding of the fiber, but instead pump light is delivered across the core multiple times because the core is coiled on itself like a rope. This configuration is suitable for power scaling in which many pump sources are used around the periphery of the coil. Fiber disk lasers have exceptional protection against back reflection compared to traditional fiber lasers. Fiber disk lasers can be used for welding and cutting applications requiring more than 1000 watts of power.

3 fiber disk lasers

References

- Wade, R. C. (1998). "Chemical Lasers with Annular Gain Media". In Kossowsky, R.; Jelinek, M.; Novák, J. Optical Resonators - Science and Engineering. Kluwer Academic. pp. 211–223. ISBN 978-0-7923-4962-4.

- Costela A, Garcia-Moreno I, Gomez C (2016). "Medical Applications of Organic Dye Lasers". In Duarte FJ. Tunable Laser Applications (3rd ed.). Boca Raton: CRC Press. pp. 293–313. ISBN 9781482261066.

- Duarte FJ, ed. (2016). Tunable Laser Applications (3rd ed.). Boca Raton: CRC Press. ISBN 9781482261066.

- Ueda; Sekiguchi H.; Matsuoka Y.; Miyajima H.; H.Kan (1999). "Conceptual design of kW-class fiber-embedded disk and tube lasers". Lasers and Electro-Optics Society 1999 12th Annual Meeting. LEOS '99. IEEE. 2: 217–218. doi:10.1109/CLEOPR.1999.811381. ISBN 0-7803-5661-6.

- "Super-bright, fast X-ray free-electron lasers can now image single layer of proteins". KurzweilAI. doi:10.1107/S2052252514001444. Retrieved 2014-02-17.

- "New Era of Research Begins as World's First Hard X-ray Laser Achieves "First Light"". SLAC National Accelerator Laboratory. April 21, 2009. Retrieved 2013-11-06

- ""Self-seeding" promises to speed discoveries, add new scientific capabilities". SLAC National Accelerator Laboratory. August 13, 2012. Retrieved 2013-11-06.

- Thomsen, Dietrich E. (1985, December 14). Strategic defense of X-ray initiative. The Free Library. (1985). Retrieved May 08, 2013

- Patel, A.; Lincoln, B.; Stone, D. (April 1, 2013). "Specialty Fiber: Fiber lasers lower cost of making SAW's". Laser Focus World. 49 (4): 59. Retrieved June 18, 2013.

3

Techniques of Laser

There are several techniques by which lasers operate to produce light. This chapter describes in detail the techniques of Q-switching, mode-locking and gain-switching. This content also includes manufacturing and marking techniques like laser bonding, laser cutting, direct metal laser sintering and laser diffraction analysis. These techniques help in understanding the way in which lasers operate and also the uses of these techniques.

Q-Switching

Q-switching, sometimes known as giant pulse formation or Q-spoiling, is a technique by which a laser can be made to produce a pulsed output beam. The technique allows the production of light pulses with extremely high (gigawatt) peak power, much higher than would be produced by the same laser if it were operating in a continuous wave (constant output) mode. Compared to modelocking, another technique for pulse generation with lasers, Q-switching leads to much lower pulse repetition rates, much higher pulse energies, and much longer pulse durations. The two techniques are sometimes applied together.

Q-switching was first proposed in 1958 by Gordon Gould, and independently discovered and demonstrated in 1961 or 1962 by R.W. Hellwarth and F.J. McClung using electrically switched Kerr cell shutters in a ruby laser.

Principle of Q-Switching

Q-switching is achieved by putting some type of variable attenuator inside the laser's optical resonator. When the attenuator is functioning, light which leaves the gain medium does not return, and lasing cannot begin. This attenuation inside the cavity corresponds to a decrease in the *Q factor or quality factor of the* optical resonator. A high Q factor corresponds to low resonator losses per roundtrip, and vice versa. The variable attenuator is commonly called a "Q-switch", when used for this purpose.

Initially the laser medium is pumped while the Q-switch is set to prevent feedback of light into the gain medium (producing an optical resonator with low Q). This produces a population inversion, but laser operation cannot yet occur since there is no feedback from the resonator. Since the rate of stimulated emission is dependent on the amount of light entering the medium, the amount of energy stored in the gain medium increases as the medium is pumped. Due to losses from spontaneous emission and other processes, after a certain time the stored energy will reach some maximum level; the medium is said to be *gain saturated. At this point, the Q-switch device is quickly changed from low to high Q, allowing feedback and the process of* optical amplification by stimulated emission to begin. Because of the large amount of energy already stored in the gain medium, the intensity of light in the laser resonator builds up very quickly; this also causes the energy stored in the medium to be depleted almost as quickly. The net result is a short pulse of light output from the laser, known as a *giant pulse, which may have a very high peak intensity.*

There are two main types of Q-switching:

Active Q-Switching

Here, the Q-switch is an externally controlled variable attenuator. This may be a mechanical device such as a shutter, chopper wheel, or spinning mirror/prism placed inside the cavity, or (more commonly) it may be some form of modulator such as an acousto-optic device, a magneto-optic effect device or an electro-optic device — a Pockels cell or Kerr cell. The reduction of losses (increase of Q) is triggered by an external event, typically an electrical signal. The pulse repetition rate can therefore be externally controlled. Modulators generally allow a faster transition from low to high Q, and provide better control. An additional advantage of modulators is that the rejected light may be coupled out of the cavity and can be used for something else. Alternatively, when the modulator is in its low-Q state, an externally generated beam can be coupled *into the cavity through the modulator. This can be used to "seed" the cavity with a beam that has desired characteristics (such as* transverse mode or wavelength). When the Q is raised, lasing builds up from the initial seed, producing a Q-switched pulse that has characteristics inherited from the seed.

Passive Q-Switching

In this case, the Q-switch is a saturable absorber, a material whose transmission increases when the intensity of light exceeds some threshold. The material may be an ion-doped crystal like Cr:YAG, which is used for Q-switching of Nd:YAG lasers, a bleachable dye, or a passive semiconductor device. Initially, the loss of the absorber is high, but still low enough to permit some lasing once a large amount of energy is stored in the gain medium. As the laser power increases, it saturates the absorber, i.e., rapidly reduces the resonator loss, so that the power can increase even faster. Ideally, this brings the absorber into a state with low losses to allow efficient extraction of the stored energy by the laser pulse. After the pulse, the absorber recovers to its high-loss state before the gain recovers, so that the next pulse is delayed until the energy in the gain medium is fully replenished. The pulse repetition rate can only indirectly be controlled, e.g. by varying the laser's pump power and the amount of saturable absorber in the cavity. Direct control of the repetition rate can be achieved by using a pulsed pump source as well as passive Q-switching.

Variants

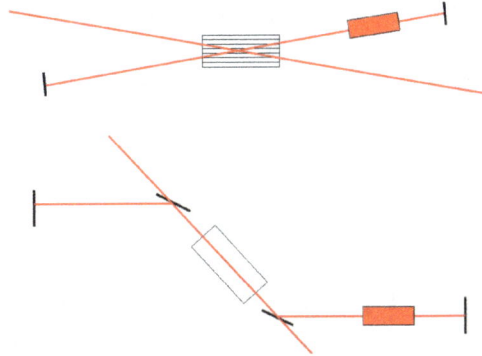

Regenerative amplifier. Red line: Laser beam. Red box: Gain medium. Top: AOM-based design. Bottom: The Pockel's cell-based design needs thin film polarizers. The direction of the emitted pulse depends on the timing.

- Jitter can be reduced by not reducing the Q by as much, so that a small amount of light can still circulate in the cavity. This provides a "seed" of light that can aid in the buildup of the next Q-switched pulse.

- Cavity dumping: The cavity end mirrors are 100% reflective, so that no output beam is produced when the Q is high. Instead, the Q-switch is used to "dump" the beam out of the cavity after a time delay. The cavity Q goes from low to high to start the laser buildup, and then goes from high to low to "dump" the beam from the cavity all at once. This produces a shorter output pulse than regular Q-switching. Electro-optic modulators are normally used for this, since they can easily be made to function as a near-perfect beam "switch" to couple the beam out of the cavity. The modulator that dumps the beam may be the same modulator that Q-switches the cavity, or a second (possibly identical) modulator. A dumped cavity is more complicated to align than simple Q-switching, and may need a control loop to choose the best time at which to dump the beam from the cavity.

- Regenerative amplification: In regenerative amplification, an optical amplifier is placed inside a Q-switched cavity. Pulses of light from another laser (the "master oscillator") are injected into the cavity by lowering the Q to allow the pulse to enter and then increasing the Q to confine the pulse to the cavity where it can be amplified by repeated passes through the gain medium. The pulse is then allowed to leave the cavity via another Q switch.

Typical Performance

A typical Q-switched laser (e.g. a Nd:YAG laser) with a resonator length of e.g. 10 cm can produce light pulses of several tens of nanoseconds duration. Even when the average power is well below 1 W, the peak power can be many kilowatts. Large-scale laser systems can produce Q-switched pulses with energies of many joules and peak powers in the gigawatt region. On the other hand, passively Q-switched microchip lasers (with very short resonators) have generated pulses with durations far below one nanosecond and pulse repetition rates from hundreds of hertz to several megahertz (MHz)

Applications

Q-switched lasers are often used in applications which demand high laser intensities in nanosecond pulses, such as metal cutting or pulsed holography. Nonlinear optics often takes advantage of the high peak powers of these lasers, offering applications such as 3D optical data storage and 3D microfabrication. However, Q-switched lasers can also be used for measurement purposes, such as for distance measurements (range finding) by measuring the time it takes for the pulse to get to some target and the reflected light to get back to the sender. It can be also used in chemical dynamic study, e.g. temperature jump relaxation study.

Q-switched lasers are also used to remove tattoos. They are used to shatter tattoo pigment into particles that are cleared by the body's lymphatic system. Full removal can take between six and twenty treatments depending on the amount and colour of ink, spaced at least a month apart, using different wavelengths for different coloured inks. Nd:YAG lasers are currently the most favoured lasers due to their high peak powers, high repetition rates and relatively low costs. In 2013 a picosecond laser was introduced based on clinical research which appears to show better clearance with 'difficult' colours such as green and light blue.

Gain-Switching

Gain-switching is a technique in optics by which a laser can be made to produce pulses of light of extremely short duration, of the order of picoseconds (10^{-12} s).

In a semiconductor laser, the optical pulses are generated by injecting a large number of carriers (electrons) into the active region of the device, bringing the carrier density within that region from below to above the lasing threshold. When the carrier density exceeds that value, the ensuing stimulated emission results in the generation of a large number of photons.

However, carriers are depleted as a result of stimulated emission faster than they are injected. So the carrier density eventually falls back to below lasing threshold which results in the termination of the optical output. If carrier injection has not ceased during this period, then the carrier density in the active region can increase once more and the process will repeat itself.

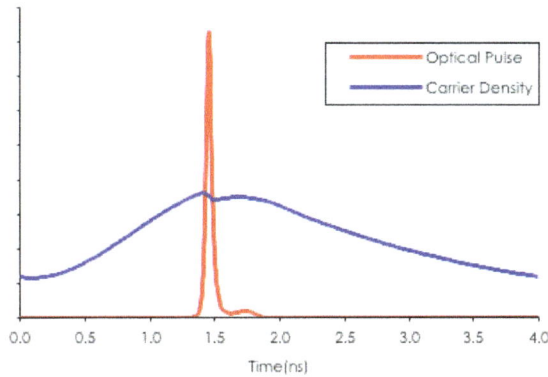

The figure on the right shows a typical pulse generated by gain-switching with a sinusoidal injection current at 250 MHz producing a pulse of approximately 50 ps. The carrier density is depleted during the pulse, and subsequently rises due to continued current injection, producing a smaller secondary pulse. If the injection current is rapidly switched off at the proper time, for example using a step recovery diode circuit, a single 50 ps light pulse can be generated.

For solid-state and dye lasers, gain switching (or *synchronous pumping*) *usually involves the laser gain medium being* pumped with another pulsed laser. Since the pump pulses are of short duration, optical gain is only present in the laser for a short time, which results in a pulsed output. Q-switching is more commonly used for producing pulsed output from these types of laser, as pulses with much higher peak power can be achieved.

The term gain-switching derives from the fact that the optical gain is negative when carrier density or pump intensity in the active region of the device is below threshold, and switches to a positive value when carrier density or the pump intensity exceeds the lasing threshold.

Mode-Locking

Mode-locking is a technique in optics by which a laser can be made to produce pulses of light of extremely short duration, on the order of picoseconds (10−12 s) or femtoseconds (10−15 s).

The basis of the technique is to induce a fixed-phase relationship between the longitudinal modes of the laser's resonant cavity. The laser is then said to be 'phase-locked' or 'mode-locked'. Interference between these modes causes the laser light to be produced as a train of pulses. Depending on the properties of the laser, these pulses may be of extremely brief duration, as short as a few femtoseconds.

Laser Cavity Modes

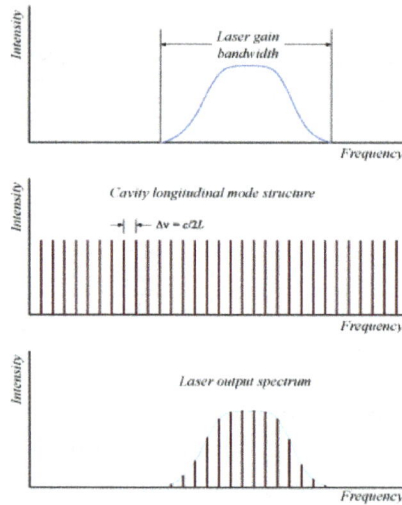

Laser mode structure

Although laser light is perhaps the purest form of light, it is not of a single, pure frequency or wavelength. All lasers produce light over some natural bandwidth or range of frequencies. A laser's bandwidth of operation is determined primarily by the gain medium from which the laser is constructed, and the range of frequencies over which a laser may operate is known as the gain bandwidth. For example, a typical helium–neon laser has a gain bandwidth of about 1.5 GHz (a wavelength range of about 0.002 nm at a central wavelength of 633 nm), whereas a titanium-doped sapphire (Ti:sapphire) solid-state laser has a bandwidth of about 128 THz (a 300-nm wavelength range centered around 800 nm).

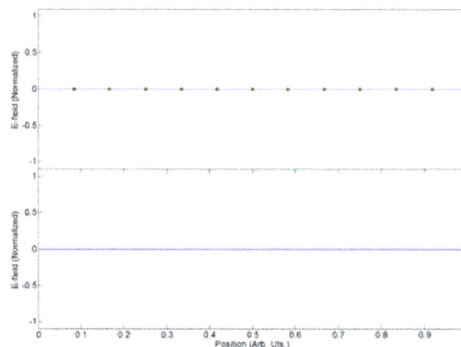

A mode-locked, fully reflecting cavity supporting the first 30 modes. The upper plot shows the first 8 modes inside the cavity (lines) and the total electric field at various positions inside the cavity (points). The lower plot shows the total electric field inside the cavity.

The second factor to determine a laser's emission frequencies is the optical cavity (or resonant cavity) of the laser. In the simplest case, this consists of two plane (flat) mirrors facing each other, surrounding the gain medium of the laser (this arrangement is known as a Fabry–Pérot cavity). Since light is a wave, when bouncing between the mirrors of the cavity, the light will constructively and destructively interfere with itself, leading to the formation of standing waves or modes between the mirrors. These standing waves form a discrete set of frequencies, known as the *longitudinal modes of the cavity. These modes are the only frequencies of light which are self-regenerating and allowed to oscillate by the resonant cavity; all other frequencies of light are suppressed by destructive interference. For a simple plane-mirror cavity, the allowed modes are those for which the separation distance of the mirrors L is an exact multiple of half the wavelength of the light λ, such that L = qλ/2, where q is an integer known as the mode order.*

In practice, L is usually much greater than λ, so the relevant values of q are large (around 10^5 to 10^6). Of more interest is the frequency separation between any two adjacent modes q and $q+1$; this is given (for an empty linear resonator of length L) by Δv:

$$\Delta v = \frac{c}{2L}$$

where *c is the* speed of light ($\approx 3 \times 10^8$ m·s−1).

Using the above equation, a small laser with a mirror separation of 30 cm has a frequency separation between longitudinal modes of 0.5 GHz. Thus for the two lasers referenced above, with a 30-cm cavity, the 1.5 GHz bandwidth of the HeNe laser would support up to 3 longitudinal modes, whereas the 128 THz bandwidth of the Ti:sapphire laser could support approximately 250,000 modes. When more than one longitudinal mode is excited, the laser is said to be in "multi-mode" operation. When only one longitudinal mode is excited, the laser is said to be in "single-mode" operation.

Each individual longitudinal mode has some bandwidth or narrow range of frequencies over which it operates, but typically this bandwidth, determined by the Q factor of the cavity, is much smaller than the intermode frequency separation.

Mode-Locking Theory

In a simple laser, each of these modes oscillates independently, with no fixed relationship between each other, in essence like a set of independent lasers all emitting light at slightly different frequencies. The individual phase of the light waves in each mode is not fixed, and may vary randomly due to such things as thermal changes in materials of the laser. In lasers with only a few oscillating modes, interference between the modes can cause beating effects in the laser output, leading to fluctuations in intensity; in lasers with many thousands of modes, these interference effects tend to average to a near-constant output intensity.

If instead of oscillating independently, each mode operates with a fixed phase between it and the other modes, the laser output behaves quite differently. Instead of a random or constant output intensity, the modes of the laser will periodically all constructively interfere with one another, producing an intense burst or pulse of light. Such a laser is said to be 'mode-locked' or 'phase-locked'. These pulses occur separated in time by $\tau = 2L/c$, where τ is the time taken for the light to make exactly one round trip of the laser cavity. This time corresponds to a frequency exactly equal to the mode spacing of the laser, $\Delta v = 1/\tau$.

The duration of each pulse of light is determined by the number of modes which are oscillating in phase (in a real laser, it is not necessarily true that all of the laser›s modes will be phase-locked). If there are N modes locked with a frequency separation Δv, the overall mode-locked bandwidth is $N\Delta v$, and the wider this bandwidth, the shorter the pulse duration from the laser. In practice, the actual pulse duration is determined by the shape of each pulse, which is in turn determined by the exact amplitude and phase relationship of each longitudinal mode. For example, for a laser producing pulses with a Gaussian temporal shape, the minimum possible pulse duration Δt is given by

$$\Delta t = \frac{0.441}{N\Delta v}.$$

The value 0.441 is known as the 'time-bandwidth product' of the pulse, and varies depending on the pulse shape. For ultrashort pulse lasers, a hyperbolic-secant-squared ($sech^2$) pulse shape is often assumed, giving a time-bandwidth product of 0.315.

Using this equation, the minimum pulse duration can be calculated consistent with the measured laser spectral width. For the HeNe laser with a 1.5-GHz spectral width, the shortest Gaussian pulse consistent with this spectral width would be around 300 picoseconds; for the 128-THz bandwidth Ti:sapphire laser, this spectral width would be only 3.4 femtoseconds. These values represent the shortest possible Gaussian pulses consistent with the laser's linewidth; in a real mode-locked laser, the actual pulse duration depends on many other factors, such as the actual pulse shape, and the overall dispersion of the cavity.

Subsequent modulation could in principle shorten the pulse width of such a laser further; however, the measured spectral width would then be correspondingly increased.

Mode-Locking Methods

Methods for producing mode-locking in a laser may be classified as either 'active' or 'passive'. Active methods typically involve using an external signal to induce a modulation of the intracavity light. Passive methods do not use an external signal, but rely on placing some element into the laser cavity which causes self-modulation of the light.

Active Mode-Locking

The most common active mode-locking technique places a standing wave acousto-optic modulator into the laser cavity. When driven with an electrical signal, this produces a sinusoidal amplitude modulation of the light in the cavity. Considering this in the frequency domain, *if a mode has optical frequency v, and is amplitude-modulated at a frequency f, the resulting signal has* sidebands at optical frequencies $v - f$ and $v + f$. *If the modulator is driven at the same frequency as the cavity-mode spacing Δv, then these sidebands correspond to the two cavity modes adjacent to the original mode. Since the sidebands are driven in-phase, the central mode and the adjacent modes will be phase-locked together. Further operation of the modulator on the sidebands produces phase-locking of the $v - 2f$ and $v + 2f$ modes, and so on until all modes in the gain bandwidth are locked. As said above, typical lasers are multi-mode and not seeded by a root mode. So multiple modes need to work out which phase to use. In a passive cavity with this locking applied there is no way to dump the* entropy given by the original independent phases. This locking is better described as a coupling, leading to a complicated behavior and not clean pulses. The coupling is only dissipative because of the dissipative nature of the amplitude modulation. Otherwise, the phase modulation would not work.

This process can also be considered in the time domain. The amplitude modulator acts as a weak ‹shutter› to the light bouncing between the mirrors of the cavity, attenuating the light when it is «closed», and letting it through when it is «open». *If the modulation rate f is synchronised to the cavity round-trip time τ, then a single pulse of light will bounce back and forth in the cavity. The actual strength of the modulation does not have to be large; a modulator that attenuates 1% of the light when «closed» will mode-lock a laser, since the same part of the light is repeatedly attenuated as it traverses the cavity.*

Related to this amplitude modulation (AM), active mode-locking is frequency modulation (FM) mode-locking, which uses a modulator device based on the electro-optic effect. This device, when placed in a laser cavity and driven with an electrical signal, induces a small, sinusoidally varying frequency shift in the light passing through it. If the frequency of modulation is matched to the round-trip time of the cavity, then some light in the cavity sees repeated upshifts in frequency, and some repeated downshifts. After many repetitions, the upshifted and downshifted light is swept out of the gain bandwidth of the laser. The only light which is unaffected is that which passes through the modulator when the induced frequency shift is zero, which forms a narrow pulse of light.

The third method of active mode-locking is synchronous mode-locking, or synchronous pumping. In this, the pump source (energy source) for the laser is itself modulated, effectively turning the laser on and off to produce pulses. Typically, the pump source is itself another mode-locked laser. This technique requires accurately matching the cavity lengths of the pump laser and the driven laser.

Passive Mode-Locking

Passive mode-locking techniques are those that do not require a signal external to the laser (such as the driving signal of a modulator) to produce pulses. Rather, they use the light in the cavity to cause a change in some intracavity element, which will then itself produce a change in the intracavity light. A commonly used device to achieve this is a saturable absorber.

A saturable absorber is an optical device that exhibits an intensity-dependent transmission. What this means is that the device behaves differently depending on the intensity of the light passing through it. For passive mode-locking, ideally a saturable absorber will selectively absorb low-intensity light, and transmit light which is of sufficiently high intensity. When placed in a laser cavity, a saturable absorber will attenuate low-intensity constant wave light (pulse wings). However, because of the somewhat random intensity fluctuations experienced by an un-mode-locked laser, any random, intense spike will be transmitted preferentially by the saturable absorber. As the light in the cavity oscillates, this process repeats, leading to the selective amplification of the high-intensity spikes, and the absorption of the low-intensity light. After many round trips, this leads to a train of pulses and mode-locking of the laser.

Considering this in the frequency domain, if a mode has optical frequency v, and is amplitude-modulated at a frequency nf, the resulting signal has sidebands at optical frequencies $v - nf$ and $v + nf$ and enables much stronger mode-locking for shorter pulses and more stability than active mode-locking, but has startup problems.

Saturable absorbers are commonly liquid organic dyes, but they can also be made from doped crystals and semiconductors. Semiconductor absorbers tend to exhibit very fast response times (~100 fs), which is one of the factors that determines the final duration of the pulses in a passively mode-locked laser. In a *colliding-pulse mode-locked laser the absorber steepens the leading edge while the lasing medium steepens the trailing edge of the pulse.*

In particular, graphene can be saturated over the visible to near-infrared region and it has a smaller non-saturable loss and higher damage threshold, compared with SWCNTs.

There are also passive mode-locking schemes that do not rely on materials that directly display an intensity dependent absorption. In these methods, nonlinear optical effects in intracavity components are used to provide a method of selectively amplifying high-intensity light in the cavity, and attenuation of low-intensity light. One of the most successful schemes is called Kerr-lens mode-locking (KLM), also sometimes called "self mode-locking". This uses a nonlinear optical process, the optical Kerr effect, which results in high-intensity light being focussed differently from low-intensity light. By careful arrangement of an aperture in the laser cavity, this effect can be exploited to produce the equivalent of an ultra-fast response time saturable absorber.

Hybrid Mode-Locking

In some semiconductor lasers a combination of the two above techniques can be used. Using a laser with a saturable absorber, and modulating the electrical injection at the same frequency the laser is locked at, the laser can be stabilized by the electrical injection. This has the advantage of stabilizing the phase noise of the laser, and can reduce the timing jitter of the pulses from the laser.

Fourier domain Mode Locking

Fourier domain mode locking (FDML) is a laser modelocking technique that creates a continuous wave, wavelength-swept light output. A main application for FDML lasers is optical coherence tomography.

Practical Mode-Locked Lasers

In practice, a number of design considerations affect the performance of a mode-locked laser. The most important are the overall dispersion of the laser's optical resonator, which can be controlled with a prism compressor or some dispersive mirrors placed in the cavity, and optical nonlinearities. For excessive net group delay dispersion (GDD) of the laser cavity, the phase of the cavity modes can not be locked over a large bandwidth, and it will be difficult to obtain very short pulses. For a suitable combination of negative (anomalous) net GDD with the Kerr nonlinearity, soliton-like interactions may stabilize the mode-locking and help to generate shorter pulses. The shortest possible pulse duration is usually accomplished either for zero dispersion (without nonlinearities) or for some slightly negative (anomalous) dispersion (exploiting the soliton mechanism).

The shortest directly produced optical pulses are generally produced by Kerr-lens mode-locked Ti-sapphire lasers, and are around 5 femtoseconds long. Alternatively, amplified pulses of a similar duration are created through the compression of longer (e.g. 30 fs) pulses via self-phase modulation in a hollow core fibre or during filamentation. However, the minimum pulse duration is limited by the period of the carrier frequency (which is about 2.7 fs for Ti:S systems), therefore shorter pulses require moving to shorter wavelengths. Some advanced techniques (involving high harmonic generation with amplified femtosecond laser pulses) can be used to produce optical features with durations as short as 100 attoseconds in the extreme ultraviolet spectral region (i.e. <30 nm). Other achievements, important particularly for laser applications, concern the development of mode-locked lasers which can be pumped with laser diodes, can generate very high average output powers (tens of watts) in sub-picosecond pulses, or generate pulse trains with extremely high repetition rates of many GHz.

Pulse durations less than approximately 100 fs are too short to be directly measured using optoelectronic techniques (i.e. photodiodes), and so indirect methods such as autocorrelation, frequency-resolved optical gating, spectral phase interferometry for direct electric-field reconstruction or multiphoton intrapulse interference phase scan are used.

Applications

- *Nuclear fusion.* (inertial confinement fusion).

- *Nonlinear optics, such as* second-harmonic generation, parametric down-conversion, optical parametric oscillators, and generation of Terahertz radiation

- *Optical Data Storage uses lasers, and the emerging technology of* 3D optical data storage generally relies on nonlinear photochemistry. For this reason, many examples use mode-locked lasers, since they can offer a very high repetition rate of ultrashort pulses.

- Femtosecond laser nanomachining – The short pulses can be used to nanomachine in many types of materials.

- An example of pico- and femtosecond micromachining is drilling the silicon jet surface of ink jet printers

- Two-photon microscopy

- *Corneal Surgery. Femtosecond lasers can create bubbles in the* cornea, if multiple bubbles are created in a planar fashion parallel to the corneal surface then the tissue separates at this plane and a flap like the one in LASIK is formed (Intralase: Intralasik or SBK (Sub Bowman Keratomileusis) if the flap thickness is equal or less than 100 micrometres). If done in multiple layers a piece of corneal tissue between these layers can be removed (Visumax: FLEX Femtosecond Lenticle Extraction).

- A laser technique has been developed that renders the surface of metals deep black. A femtosecond laser pulse deforms the surface of the metal forming nanostructures. The immensely increased surface area can absorb virtually all the light that falls on it thus rendering it deep black. This is one type of black gold

- Photonic Sampling, using the high accuracy of lasers over electronic clocks to decrease the sampling error in electronic ADCs

Laser Bonding

Laser bonding is a marking technique that uses lasers to bond an additive marking substance to a substrate.

First invented in the mid 1990s by Paul W. Harrison, the founder of TherMark, LLC, this patent protected and patent pending technology produces permanent marks on metal, glass, ceramic and plastic parts for a diverse range of industrial and artistic applications, ranging from aerospace and medical to the awards and engraving industries. It differs from the more widely known techniques of laser engraving and laser ablation in that it is an additive process, adding material to the substrate surface instead of removing it.

Laser bonding has been achieved by Nd:YAG, CO_2 laser, Fiber laser and Diode-pumped solid-state laser and can be accomplished using other forms of radiant energy.

The Laser Bonding Process

Mark quality depends on a variety of factors, including the substrate used, marking speed, laser spot size, beam overlap, materials thickness, and laser parameters. Laser bonding materials may be applied by various methods, including a brush on technique, spraying, pad printing, screen printing, roll coating, tape, and others.

The marking process generally comprises three steps:

1. Application of the marking material.

2. Irradiating the marking material with a laser in the form of the desired mark.

3. Removal of excess, unbonded material.

The resulting marking is permanently bonded to the substrate, and in most cases it is as durable as the substrate itself.

The Durability of Laser Bonded Markings

Markings placed on stainless steel are extremely durable and have survived such testing as abrasion resistance, chemical resistance, outdoor exposure, extreme heat, extreme cold, acids, bases and various organic solvents.

Marks on glass have been tested for resistance to acids, bases and scratching.

NASA's International Space Station, or ISS, was home to aluminum squares laser marked with CerMark® marking material for almost four years. These squares were part of the Material International Space Station Experiment, or MISSE.

In this experiment test markings were applied to coupons made of materials commonly used in the construction of the external components used on space transportation vehicles, satellites and space stations. Markings applied using a wide range of different methods and techniques, including laser bonding. The material test coupons were then affixed to spaces provided on test panels, which were then installed onto trays which were attached to the ISS during a space walk conducted during the STS-105 Mission flown on August 10, 2001. The trays were positioned on the ISS so that they could expect to receive the maximum amount of impact damage and exposure to a high degree of atomic oxygen and UV radiation.

The experiment was recovered on July 30, 2005 during STS-114 and returned to earth on August 9, 2005. The markings, DataMatrix two dimensional bar codes, were evaluated and found to be readable and visually looked as good as the day they were placed in orbit.

The laser bonding process is outlined and specified in both military and NASA marking specifications and standards. Laser bonding is also a preferred technique for use in the United States Department of Defense "Item Unique Identification" system (IUID).

Laser Cutting

Laser cutting is a technology that uses a laser to cut materials, and is typically used for industrial manufacturing applications, but is also starting to be used by schools, small businesses, and hobbyists. Laser cutting works by directing the output of a high-power laser most commonly through optics. The laser optics and CNC (computer numerical control) are used to direct the material or the laser beam generated. A typical commercial laser for cutting materials would involve a motion control system to follow a CNC or G-code of the pattern to be cut onto the material. The focused laser beam is directed at the material, which then either melts, burns, vaporizes away, or is blown away by a jet of gas, leaving an edge with a high-quality surface finish. Industrial laser cutters are used to cut flat-sheet material as well as structural and piping materials.

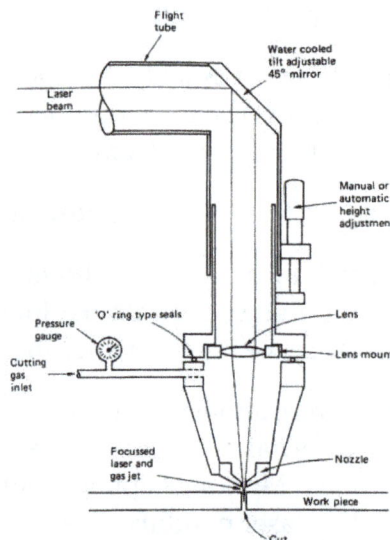

Diagram of a laser cutter

Laser cutting process on a sheet of steel

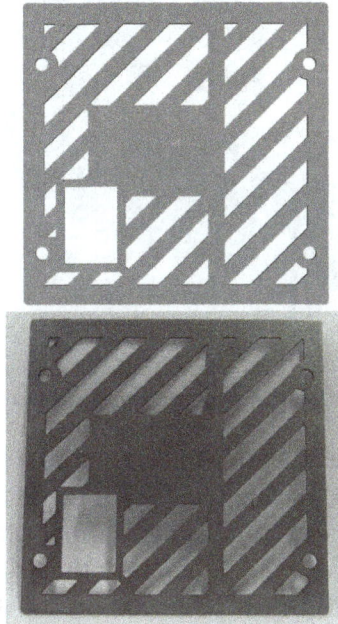

CAD (top) and stainless steel laser-cut part (bottom)

History

In 1965, the first production laser cutting machine was used to drill holes in diamond dies. This machine was made by the Western Electric Engineering Research Center. In 1967, the British pioneered laser-assisted oxygen jet cutting for metals. In the early 1970s, this technology was put into production to cut titanium for aerospace applications. At the same time CO_2 lasers were adapted to cut non-metals, such as textiles, because, at the time, CO_2 lasers were not powerful enough to overcome the thermal conductivity of metals.

Process

Generation of the laser beam involves stimulating a lasing material by electrical discharges or lamps within a closed container. As the lasing material is stimulated, the beam is reflected internally by means of a partial mirror, until it achieves sufficient energy to escape as a stream of monochromatic coherent light. Mirrors or fiber optics are typically used to direct the coherent light to a lens, which focuses the light at the work zone. The narrowest part of the focused beam is generally less than 0.0125 inches (0.32 mm). in diameter. Depending upon material thickness, kerf widths as small as 0.004 inches (0.10 mm) are possible. In order to be able to start cutting from somewhere other than the edge, a pierce is done before every cut. Piercing usually involves a high-power pulsed laser beam which slowly makes a hole in the material, taking around 5–15 seconds for 0.5-inch-thick (13 mm) stainless steel, for example.

Industrial laser cutting of steel with cutting instructions programmed through the CNC interface

The parallel rays of coherent light from the laser source often fall in the range between 0.06–0.08 inches (1.5–2.0 mm) in diameter. This beam is normally focused and intensified by a lens or a mirror to a very small spot of about 0.001 inches (0.025 mm) to create a very intense laser beam. In order to achieve the smoothest possible finish during contour cutting, the direction of beam polarization must be rotated as it goes around the periphery of a contoured workpiece. For sheet metal cutting, the focal length is usually 1.5–3 inches (38–76 mm).

Advantages of laser cutting over mechanical cutting include easier workholding and reduced contamination of workpiece (since there is no cutting edge which can become contaminated by the material or contaminate the material). Precision may be better, since the laser beam does not wear during the process. There is also a reduced chance of warping the material that is being cut, as laser systems have a small heat-affected zone. Some materials are also very difficult or impossible to cut by more traditional means.

Laser cutting for metals has the advantages over plasma cutting of being more precise and using less energy when cutting sheet metal; however, most industrial lasers cannot cut through the greater metal thickness that plasma can. Newer lasers machines operating at higher power (6000 watts, as contrasted with early laser cutting machines' 1500 watt ratings) are approaching plasma machines in their ability to cut through thick materials, but the capital cost of such machines is much higher than that of plasma cutting machines capable of cutting thick materials like steel plate.

Types

A diffusion cooled resonator

4000 watt CO2 laser cutter

There are three main types of lasers used in laser cutting. The CO2 laser is suited for cutting, boring, and engraving. The neodymium (Nd) and neodymium yttrium-aluminium-garnet (Nd-YAG) lasers are identical in style and differ only in application. Nd is used for boring and where high energy but low repetition are required. The Nd-YAG laser is used where very high power is needed and for boring and engraving. Both CO2 and Nd/ Nd-YAG lasers can be used for welding.

Common variants of CO2 lasers include fast axial flow, slow axial flow, transverse flow, and slab.

CO2 lasers are commonly "pumped" by passing a current through the gas mix (DC-excited) or using radio frequency energy (RF-excited). The RF method is newer and has become more popular. Since DC designs require electrodes inside the cavity, they can encounter electrode erosion and plating of electrode material on glassware and optics. Since RF resonators have external electrodes they are not prone to those problems.

CO2 lasers are used for industrial cutting of many materials including mild steel, aluminium, stainless steel, titanium, Taskboard, paper, wax, plastics, wood, and fabrics. YAG lasers are primarily used for cutting and scribing metals and ceramics.

In addition to the power source, the type of gas flow can affect performance as well. In a fast axial flow resonator, the mixture of carbon dioxide, helium and nitrogen is circulated at high velocity by a turbine or blower. Transverse flow lasers circulate the gas mix at a lower velocity, requiring a simpler blower. Slab or diffusion cooled resonators have a static gas field that requires no pressurization or glassware, leading to savings on replacement turbines and glassware.

The laser generator and external optics (including the focus lens) require cooling. Depending on system size and configuration, waste heat may be transferred by a coolant or directly to air. Water is a commonly used coolant, usually circulated through a chiller or heat transfer system.

A *laser microjet is a water-jet guided* laser in which a pulsed laser beam is coupled into a low-pressure water jet. This is used to perform laser cutting functions while using the water jet to guide the laser beam, much like an optical fiber, through total internal reflection. The advantages of this are that the water also removes debris and cools the material. Additional advantages over traditional "dry" laser cutting are high dicing speeds, parallel kerf, and omnidirectional cutting.

Fiber lasers are a type of solid state laser that is rapidly growing within the metal cutting industry. Unlike CO2, Fiber technology utilizes a solid gain medium, as opposed to a gas or liquid. The "seed laser" produces the laser beam and is then amplified within a glass fiber. With a wavelength of only 1.064 micrometers fiber lasers produce an extremely small spot size (up to 100 times smaller compared to the CO2) making it ideal for cutting reflective metal material. This is one of the main advantages of Fiber compared to CO2.

Methods

There are many different methods in cutting using lasers, with different types used to cut different material. Some of the methods are vaporization, melt and blow, melt blow and burn, thermal stress cracking, scribing, cold cutting and burning stabilized laser cutting.

Vaporization cutting

> In vaporization cutting the focused beam heats the surface of the material to boiling point and generates a keyhole. The keyhole leads to a sudden increase in absorptivity quickly deepening the hole. As the hole deepens and the material boils, vapor generated erodes the molten walls blowing ejecta out and further enlarging the hole. Non melting material such as wood, carbon and thermoset plastics are usually cut by this method.

Melt and blow

> Melt and blow or fusion cutting uses high-pressure gas to blow molten material

from the cutting area, greatly decreasing the power requirement. First the material is heated to melting point then a gas jet blows the molten material out of the kerf avoiding the need to raise the temperature of the material any further. Materials cut with this process are usually metals.

Thermal stress cracking

Brittle materials are particularly sensitive to thermal fracture, a feature exploited in thermal stress cracking. A beam is focused on the surface causing localized heating and thermal expansion. This results in a crack that can then be guided by moving the beam. The crack can be moved in order of m/s. It is usually used in cutting of glass.

Stealth dicing of silicon wafers

The separation of microelectronic chips as prepared in semiconductor device fabrication from silicon wafers may be performed by the so-called stealth dicing process, which operates with a pulsed Nd:YAG laser, the wavelength of which (1064 nm) is well adopted to the electronic band gap of silicon (1.11 eV or 1117 nm).

Reactive cutting

Also called "burning stabilized laser gas cutting", "flame cutting". Reactive cutting is like oxygen torch cutting but with a laser beam as the ignition source. Mostly used for cutting carbon steel in thicknesses over 1 mm. This process can be used to cut very thick steel plates with relatively little laser power.

Tolerances and Surface Finish

New laser cutters have positioning accuracy of 10 micrometers and repeatability of 5 micrometers.

Standard roughness Rz increases with the sheet thickness, but decreases with laser power and cutting speed. When cutting low carbon steel with laser power of 800 W, standard roughness Rz is 10 µm for sheet thickness of 1 mm, 20 µm for 3 mm, and 25 µm for 6 mm. $Rz = 12.528 \cdot (S^{0.542}) / ((P^{0.528}) \cdot (V^{0.322}))$, where: $S =$ steel sheet thickness in mm; $P =$ laser power in kW (some new laser cutters have laser power of 4 kW.); $V =$ cutting speed in meters per minute.

This process is capable of holding quite close tolerances, often to within 0.001 inch (0.025 mm) Part geometry and the mechanical soundness of the machine have much to do with tolerance capabilities. The typical surface finish resulting from laser beam cutting may range from 125 to 250 micro-inches (0.003 mm to 0.006 mm).

Machine Configurations

Dual-pallet flying optics laser

Flying optics laser head

There are generally three different configurations of industrial laser cutting machines: moving material, hybrid, and flying optics systems. These refer to the way that the laser beam is moved over the material to be cut or processed. For all of these, the axes of motion are typically designated X and Y axis. If the cutting head may be controlled, it is designated as the Z-axis.

Moving material lasers have a stationary cutting head and move the material under it. This method provides a constant distance from the laser generator to the workpiece and a single point from which to remove cutting effluent. It requires fewer optics, but requires moving the workpiece. This style machine tends to have the fewest beam delivery optics, but also tends to be the slowest.

Hybrid lasers provide a table which moves in one axis (usually the X-axis) and move the head along the shorter (Y) axis. This results in a more constant beam delivery path length than a flying optic machine and may permit a simpler beam delivery system. This can result in reduced power loss in the delivery system and more capacity per watt than flying optics machines.

Flying optics lasers feature a stationary table and a cutting head (with laser beam) that moves over the workpiece in both of the horizontal dimensions. Flying optics cutters keep the workpiece stationary during processing and often do not require material clamping. The moving mass is constant, so dynamics are not affected by varying size of the workpiece. Flying optics machines are the fastest type, which is advantageous when cutting thinner workpieces.

Flying optic machines must use some method to take into account the changing beam length from near field (close to resonator) cutting to far field (far away from resonator) cutting. Common methods for controlling this include collimation, adaptive optics or the use of a constant beam length axis.

five and six-axis machines also permit cutting formed workpieces. In addition, there are various methods of orienting the laser beam to a shaped workpiece, maintaining a proper focus distance and nozzle standoff, etc.

Pulsing

Pulsed lasers which provide a high-power burst of energy for a short period are very effective in some laser cutting processes, particularly for piercing, or when very small holes or very low cutting speeds are required, since if a constant laser beam were used, the heat could reach the point of melting the whole piece being cut.

Most industrial lasers have the ability to pulse or cut CW (Continuous Wave) under NC (numerical control) program control.

Double pulse lasers use a series of pulse pairs to improve material removal rate and hole quality. Essentially, the first pulse removes material from the surface and the second prevents the ejecta from adhering to the side of the hole or cut.

Power Consumption

The main disadvantage of laser cutting is the high power consumption. Industrial laser efficiency may range from 5% to 45%. The power consumption and efficiency of any particular laser will vary depending on output power and operating parameters. This will depend on type of laser and how well the laser is matched to the work at hand. The amount of laser cutting power required, known as *heat input, for a particular job depends on the material type, thickness, process (reactive/inert) used, and desired cutting rate.*

Amount of heat input required for various material at various thicknesses using a CO_2 laser [watts]					
Material	Material thickness				
	0.51 mm	1.0 mm	2.0 mm	3.2 mm	6.4 mm
Stainless steel	1000	1000	1000	1500	2500
Aluminium	1000	1000	1000	3800	10000
Mild steel	–	400	–	500	–
Titanium	250	210	210	–	-
Plywood	–	-	–	-	650
Boron/epoxy	–	-	–	3000	–

Production and Cutting Rates

The maximum cutting rate (production rate) is limited by a number of factors including laser power, material thickness, process type (reactive or inert,) and material properties. Common industrial systems (≥ 1 kW) will cut carbon steel metal from $0.51 - 13$ mm in thickness. For all intents and purposes, a laser can be up to thirty times faster than standard sawing.

Cutting rates using a CO_2 laser [cm/second]						
Workpiece material	Material thickness					
	0.51 mm	1.0 mm	2.0 mm	3.2 mm	6.4 mm	13 mm
Stainless steel	42.3	23.28	13.76	7.83	3.4	0.76
Aluminium	33.87	14.82	6.35	4.23	1.69	1.27
Mild steel	–	8.89	7.83	6.35	4.23	2.1
Titanium	12.7	12.7	4.23	3.4	2.5	1.7
Plywood	–	-	–	-	7.62	1.9
Boron / epoxy	–	-	–	2.5	2.5	1.1

Direct Metal Laser Sintering

Direct metal laser sintering (DMLS) is an additive manufacturing technique that uses a Yb (Ytterbium) fibre laser fired into a bed of powdered metal, aiming the laser automatically at points in space defined by a 3D model, melting or rather, welding the material together to create a solid structure. DMLS was developed by Elextrolux Rapid Development and the EOS firm of Munich, Germany.

The DMLS process involves use of a 3D CAD model whereby a .stl file is created and sent to the machine's software. A technician works with this 3D model to properly orient the geometry for part building and adds supports structure as appropriate. Once this "build file" has been completed, it is "sliced" into the layer thickness the machine will build in and downloaded to the DMLS machine allowing the build to begin. The DMLS machine uses a high-powered 200 watt Yb-fiber optic laser. Inside the build chamber area, there is a material dispensing platform and a build platform along with a recoater blade used to move new powder over the build platform. The technology fuses metal powder into a solid part by melting it locally using the focused laser beam. Parts are built up additively layer by layer, typically using layers 20 micrometers thick. This process allows for highly complex geometries to be created directly from the 3D CAD data, fully automatically, in a relatively short time and without any tooling. DMLS is a net-shape process, producing parts with high accuracy and detail resolution, good surface quality and excellent mechanical properties.

Benefits

DMLS has many benefits over traditional manufacturing techniques. The ability to quickly produce a unique part is the most obvious because no special tooling is required and parts can be built in a matter of hours. Additionally, DMLS allows for more rigorous testing of prototypes. Since DMLS can use most alloys, prototypes can now be functional hardware made out of the same material as production components.

DMLS is also one of the few additive manufacturing technologies being used in production. Since the components are built layer by layer, it is possible to design internal features and passages that could not be cast or otherwise machined. Complex geometries and assemblies with multiple components can be simplified to fewer parts with a more cost effective assembly. DMLS does not require special tooling like castings, so it is convenient for short production runs.

Applications

This technology is used to manufacture direct parts for a variety of industries including aerospace, dental, medical and other industries that have small to medium size, highly complex parts and the tooling industry to make direct tooling inserts. With a typical build envelope (e.g. for EOS's EOSINT M280) of 250 x 250 x 325 mm, and the ability to 'grow' multiple parts at one time, DMLS is a very cost and time effective technology. The technology is used both for rapid prototyping, as it decreases development time for new products, and production manufacturing as a cost saving method to simplify assemblies and complex geometries.

The Northwestern Polytechnical University of China is using a similar system to build structural titanium parts for aircraft. An EADS study shows that use of the process would reduce materials and waste in aerospace applications.

On September 5, 2013 Elon Musk tweeted an image of SpaceX's regeneratively-cooled SuperDraco rocket engine chamber emerging from an EOS 3D metal printer, noting that it was composed of the Inconel superalloy. In a surprise move, SpaceX announced in May 2014 that the flight-qualified version of the SuperDraco engine is fully printed, and is the first fully printed rocket engine. Using Inconel, an alloy of nickel and iron, additively-manufactured by direct metal laser sintering, the engine operates at a chamber pressure of 6,900 kilopascals (1,000 psi) at a very high temperature. The engines are contained in a printed protective nacelle, also DMLS-printed, to prevent fault propagation in the event of an engine failure. The engine completed a full qualification test in May 2014, and is slated to make its first orbital spaceflight in May 2017.

The ability to 3D print the complex parts was key to achieving the low-mass objective of the engine. According to Elon Musk, "It's a very complex engine, and it was very difficult to form all the cooling channels, the injector head, and the throttling mechanism. Being able to print very high strength advanced alloys … was crucial to being able to create the SuperDraco engine as it is." The 3D printing process for the SuperDraco engine dramatically reduces lead-time compared to the traditional cast parts, and "has superior strength, ductility, and fracture resistance, with a lower variability in materials properties."

Constraints

The aspects of size, feature details and surface finish, as well as print through error[*clarification needed*] in the Z axis may be factors that should be considered prior to the use of the technology.[*according to whom?*] However, by planning the build in the machine where most features are built in the x and y axis as the material is laid down, the feature tolerances can be managed well. Surfaces usually have to be polished to achieve mirror or extremely smooth finishes.

For production tooling, material density of a finished part or insert should be addressed prior to use.[*according to whom?*] For example, in injection molding inserts, any surface imperfections will cause imperfections in the plastic part, and the inserts will have to mate with the base of the mold with temperature and surfaces to prevent problems.

Independent of the material system used, the DMLS process leaves a grainy surface finish due to "powder particle size, layer-wise building sequence and [the spreading of the metal powder prior to sintering by the powder distribution mechanism]."

Metallic support structure removal and post processing of the part generated may be a time consuming process and require the use of machining, EDM and/or grinding machines having the same level of accuracy provided by the RP machine.

Laser polishing by means of shallow surface melting of DMLS-produced parts is able to reduce surface roughness by use of a fast-moving laser beam providing "just enough heat energy to cause melting of the surface peaks. The molten mass then flows into the surface valleys by surface tension, gravity and laser pressure, thus diminishing the roughness."

When using rapid prototyping machines, .stl files, which do not include anything but raw mesh data in binary (generated from Solid Works, CATIA, or other major CAD programs) need further conversion to .cli & .sli files (the format required for non stereolithography machines). Software converts .stl file to .sli files, as with the rest of the process, there can be costs associated with this step.

Materials

Currently available alloys used in the process include 17-4 and 15-5 stainless steel, maraging steel, cobalt chromium, inconel 625 and 718, aluminum AlSi10Mg, and titanium Ti6Al4V.

Laser Diffraction Analysis

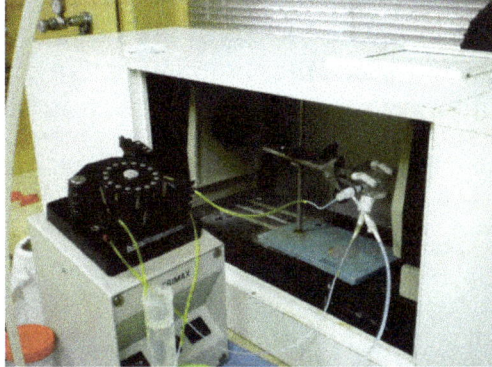

Laser diffraction analyser

Laser diffraction analysis, also known as laser diffraction spectroscopy, is a technology that utilizes diffraction patterns of a laser beam passed through any object ranging from nanometers to millimeters in size to quickly measure geometrical dimensions of a particle. This process does not depend on volumetric flow rate, the amount of particles that passes through a surface over time.

Operation

Laser diffraction analysis is based on the Fraunhofer diffraction theory, stating that the intensity of light scattered by a particle is directly proportional to the particle size. The angle of the laser beam and particle size have an inversely proportional relationship, where the laser beam angle increases as particle size decreases and vice versa.

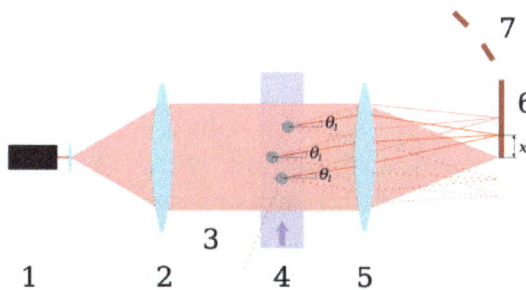

Particles moving through the spread parallel laser beam

Laser diffraction analysis is accomplished via a red He-Ne laser, a commonly used gas laser for physics experiments that is made up of a laser tube, a high-voltage power supply, and structural packaging. Alternatively, blue laser diodes or LEDs of shorter wavelength may be used. Angling of the light energy produced by the laser is detected by having a beam of light go through a suspension and then onto a sensor. A lens is placed between the object being analyzed and the detector's focal point, causing only the surrounding laser diffraction to appear. The sizes the laser can analyze depend on the lens' focal length, the distance from the lens to its point of focus. As the focal length increases, the area the laser can detect increases as well, displaying a proportional relationship. A computer can then be used to detect the object's particle sizes from the light energy produced and its layout, which the computer derives from the data collected on the particle frequencies and wavelengths.

Uses

Laser diffraction analysis has been used to measure particle-size objects in situations such as:

- observing distribution of sediments such as clay and mud, with an emphasis on silt and the sizes of bigger samples of clay.

- determining in situ measurements of particles in estuaries. Particles in estuaries are important as they allow for natural or pollutant chemical species to move around with ease. The size, density, and stability of particles in estuaries are important for their transportation. Laser diffraction analysis is used here to compare particle size distributions to support this claim as well as find cycles of change in estuaries that occur because of different particles.

- soil and its stability when wet. The stability of soil aggregation (clumps held together by moist clay) and clay dispersion (clay separating in moist soil), the two different states of soil in the Cerrado savanna region, were compared with laser diffraction analysis to determine if plowing had an effect on the two. Measurements were made before plowing and after plowing for different intervals of time. Clay dispersion turned out to not be affected by plowing while soil aggregation did.

Comparisons

Since laser diffraction analysis is not the sole way of measuring particles it has been compared to the sieve-pipette method, which is a traditional technique for grain size analysis. When compared, results showed that laser diffraction analysis made fast calculations that were easy to recreate after a one-time analysis, did not need large sample sizes, and produced large amounts of data. Results can easily be manipulated because the data is on a digital surface. Both the sieve-pipette method and laser diffraction analysis are able to analyze minuscule objects, but laser diffraction analysis resulted in having better precision than its counterpart method of particle measurement.

Criticism

Laser diffraction analysis has been questioned in validity in the following areas:

- assumptions including particles having random configurations and volume values. In some dispersion units, particles have been shown to align themselves together rather than have a turbulent flow, causing them to lead themselves in an orderly direction.

- algorithms used in laser diffraction analysis are not thoroughly validated. Different algorithms are used at times to have collected data match assumptions made by users as an attempt to avoid data that looks incorrect.

- measurement inaccuracies due to sharp edges on objects. Laser diffraction analysis has the chance of detecting imaginary particles at sharp edges because of the large angles the lasers make upon them.

- when compared to the data collecting of optical imaging, another particle-sizing technique, correlation between the two was poor for non-spherical particles. This is due to the fact that the underlying Fraunhofer and Mie theories only cover spherical particles. Non-spherical particles cause more diffuse scatter patterns and are more difficult to interpret. Some manufacturers have included algorithms in their software, which can partly compensate for non-spherical particles.

References

- Früngel, Frank B. A. (2014). Optical Pulses - Lasers - Measuring Techniques. Academic Press. p. 192. ISBN 9781483274317. Retrieved 1 February 2015.

- Taylor, Nick (2000). LASER: The inventor, the Nobel laureate, and the thirty-year patent war. New York: Simon & Schuster. ISBN 0-684-83515-0.

- e-Manufacturing Solutions. "M-Solutions Laser sintering system EOSINT M 280 for the production of tooling inserts, prototype parts and end products directly in metal" (PDF). Retrieved 2016-08-12.

- Bergin, Chris (2014-05-30). "SpaceX lifts the lid on the Dragon V2 crew spacecraft". NASAspaceflight.com. Retrieved 2015-03-06.

- Paul W. Harrison (July 2006), White Paper: "Product Identification in Automated Manufacturing" (PDF), Los Angeles, CA, retrieved 29 January 2015

- Forsman, A; et al. (June 2007). "Superpulse A nanosecond pulse format to improve laser drilling" (PDF). Photonics Spectra. Retrieved June 16, 2014.

- Kramer, Miriam (2014-05-30). "SpaceX Unveils Dragon V2 Spaceship, a Manned Space Taxi for Astronauts — Meet Dragon V2: SpaceX's Manned Space Taxi for Astronaut Trips". space.com. Retrieved 2014-05-30.

- Reiner, J. E.; Robertson, J. W. F.; Burden, D. L.; Burden, L. K.; Balijepalli, A.; Kasianowicz, J. J. (2013). "Temperature Sculpting in Yoctoliter Volumes". Journal of the American Chemical Society. doi:10.1021/ja309892e. ISSN 0002-7863.

Major Uses of Laser

There are numerous gadgets that use laser technology in their workings. This chapter examines devices like optical disc drive, laser printers, 3D scanner, barcode readers, magneto-optical trap and laser pointer and techniques like laser-based angle-resolved photoemission spectroscopy, fiber-optic communication etc. The section details the applications and design of such devices and techniques. The topics discussed in the chapter are of great importance to broaden the existing knowledge on laser.

Optical Disc Drive

A CD/DVD-ROM computer drive

In computing, an optical disc drive (ODD) is a disk drive that uses laser light or electromagnetic waves within or near the visible light spectrum as part of the process of reading or writing data to or from optical discs. Some drives can only read from certain discs, but recent drives can both read and record, also called burners or writers. Compact discs, DVDs, and Blu-ray discs are common types of optical media which can be read and recorded by such drives. Optical disc drives that are no longer in production include CD-ROM drive, CD writer drive, and combo (CD-RW/DVD-ROM) drive. As of 2015, DVD writer drive is the most common for desktop PCs and laptops. There are also the DVD-ROM drive, BD-ROM drive, Blu-ray Disc combo (BD-ROM/DVD±RW/CD-RW) drive, and Blu-ray Disc writer drive.

An external Apple USB SuperDrive

Optical disc drives are an integral part of standalone appliances such as CD players, VCD players, DVD players, Blu-ray disc players, DVD recorders, certain desktop video game consoles, such as Sony PlayStation 4, Microsoft Xbox One, Nintendo Wii U, and Sony PlayStation 3, and certain portable video game consoles, such as Sony PlayStation Portable. They are also very commonly used in computers to read software and consumer media distributed on disc, and to record discs for archival and data exchange purposes. Floppy disk drives, with capacity of 1.44 MB, have been made obsolete: optical media are cheap and have vastly higher capacity to handle the large files used since the days of floppy discs, and the vast majority of computers and much consumer entertainment hardware have optical writers. USB flash drives, high-capacity, small, and inexpensive, are suitable where read/write capability is required.

The CD/DVD drive lens on an Acer laptop

A Blu-ray drive holds independent lenses for Blu-ray and DVD/CD media. Pictured are lenses from a Blu-ray writer in a Sony Vaio E series laptop

Disc recording is restricted to storing files playable on consumer appliances (films, music, etc.), relatively small volumes of data (e.g., a standard DVD holds 4.7 gigabytes) for local use, and data for distribution, but only on a small scale; mass-producing large numbers of identical discs is cheaper and faster than individual recording.

Optical discs are used to back up relatively small volumes of data, but backing up of entire hard drives, which as of 2015[update] typically contain many hundreds of gigabytes or even multiple terabytes, is less practical. Large backups are often instead made on external hard drives, as their price has dropped to a level making this viable; in professional environments magnetic tape drives are also used.

History

The first laser disc, demonstrated in 1972, was the *Laservision* 12-inch video disc. The video signal was stored as an analog format like a video cassette. The first digitally recorded optical disc was a 5-inch audio compact disc (CD) in a read-only format created by Philips and Sony in 1975. Five years later, the same two companies introduced a digital storage solution for computers using this same CD size called a CD-ROM. Not until 1987 did Sony demonstrate the erasable and rewritable 5.25-inch optical drive.

Key Components

Laser And Optics

The most important part of an optical disc drive is an *optical path, placed in a pick-up head (PUH),* usually consisting of semiconductor laser, a lens for guiding the laser beam, and photodiodes detecting the light reflection from disc's surface.

Initially, CD lasers with a wavelength of 780 nm were used, being within infrared range. For DVDs, the wavelength was reduced to 650 nm (red color), and the wavelength for Blu-ray Disc was reduced to 405 nm (violet color).

Two main servomechanisms are used, the first one to maintain a correct distance between lens and disc, and ensure the laser beam is focused on a small *laser spot on the disc. The second servo moves a head along the disc's radius, keeping the beam on a groove, a continuous spiral data path.*

The optical sensor out of a CD/DVD drive. The two bigger rectangles are the photodiodes for pits, the inner one for land. This one also includes amplification and minor processing.

On *read only media (ROM), during the manufacturing process the groove, made of pits, is pressed on a flat surface, called land. Because the depth of the pits is approximately one-quarter to one-sixth of the laser's wavelength, the reflected beam's phase is shifted in relation to the incoming reading beam, causing mutual destructive* interference and reducing the reflected beam's intensity. This is detected by photodiodes that output electrical signals.

A recorder encodes (or *burns) data onto a recordable* CD-R, DVD-R, DVD+R, or BD-R disc (called a *blank) by selectively heating parts of an organic* dye layer with a laser. This changes the reflectivity of the dye, thereby creating marks that can be read like the pits and lands on pressed discs. For recordable discs, the process is permanent and the media can be written to only once. While the reading laser is usually not stronger than 5 mW, the writing laser is considerably more powerful. The higher the writing speed, the less time a laser has to heat a point on the media, thus its power has to increase proportionally. DVD burners' lasers often peak at about 200 mW, either in continuous wave and pulses, although some have been driven up to 400 mW before the diode fails.

For rewritable CD-RW, DVD-RW, DVD+RW, DVD-RAM, or BD-RE media, the laser is used to melt a crystalline metal alloy in the recording layer of the disc. Depending on the amount of power applied, the substance may be allowed to melt back (change the phase back) into crystalline form or left in an amorphous form, enabling marks of varying reflectivity to be created.

Double-sided media may be used, but they are not easily accessed with a standard drive, as they must be physically turned over to access the data on the other side.

Double layer (DL) media have two independent data layers separated by a semi-reflective layer. Both layers are accessible from the same side, but require the optics to change the laser's focus. Traditional single layer (SL) writable media are produced with a spiral groove molded in the protective polycarbonate layer (not in the data recording layer), to lead and synchronize the speed of recording head. Double-layered writable media have: a first polycarbonate layer with a (shallow) groove, a first data layer, a semi-reflective layer, a second (spacer) polycarbonate layer with another (deep) groove, and a second data layer. The first groove spiral usually starts on the inner edge and extends outwards, while the second groove starts on the outer edge and extends inwards.

Some drives support Hewlett-Packard's LightScribe photothermal printing technology for labeling specially coated discs.

Rotational Mechanism

The rotational mechanism in an optical drive differs considerably from that of a hard disk drive's, in that the latter keeps a constant angular velocity (CAV), in other words a constant number of revolutions per minute (RPM). With CAV, a higher throughput is generally achievable at the outer disc compared to the inner.

On the other hand, optical drives were developed with an assumption of achieving a constant throughput, in CD drives initially equal to 150 KiB/s. It was a feature important for streaming audio data that always tend to require a constant bit rate. But to ensure no disc capacity was wasted, a head had to transfer data at a maximum linear rate at all times too, without slowing on the outer rim of disc. This led to optical drives—until recently—operating with a constant linear velocity (CLV). The spiral *groove of the disc passed under its head at a constant speed. The implication of CLV, as opposed to CAV, is that disc angular velocity is no longer constant, and the spindle motor needed to be designed to vary its speed from between 200 RPM on the outer rim and 500 RPM on the inner.*

Later CD drives kept the CLV paradigm, but evolved to achieve higher rotational speeds, popularly described in multiples of a base speed. As a result, a 4× drive, for instance, would rotate at 800-2000 RPM, while transferring data steadily at 600 KiB/s, which is equal to 4 × 150 KiB/s.

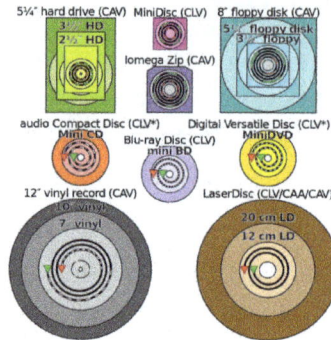

Comparison of several forms of disk storage showing tracks (not-to-scale); green denotes start and red denotes end.* Some CD-R(W) and DVD-R(W)/DVD+R(W) recorders operate in ZCLV, CAA or CAV modes.

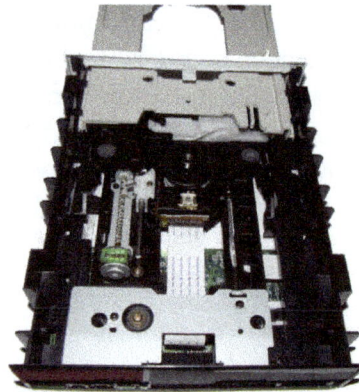

A CD-ROM Drive (without case)

For DVDs, base or 1× speed is 1.385 MB/s, equal to 1.32 MiB/s, approximately 9 times faster than the CD base speed. For Blu-ray drives, base speed is 6.74 MB/s, equal to 6.43 MiB/s.

Because keeping a constant transfer rate for the whole disc is not so important in most contemporary CD uses, a pure CLV approach had to be abandoned to keep the rotational speed of the disc safely low while maximizing data rate. Some drives work in a partial CLV (PCLV) scheme, by switching from CLV to CAV only when a rotational limit is reached. But switching to CAV requires considerable changes in hardware design, so instead most drives use the zoned constant linear velocity (Z-CLV) scheme. This divides the disc into several zones, each having its own constant linear velocity. A Z-CLV recorder rated at "52×", for example, would write at 20× on the innermost zone and then progressively increase the speed in several discrete steps up to 52× at the outer rim. Without higher rotational speeds, increased read performance may be attainable by simultaneously reading more than one point of a data groove, but drives with such mechanisms are more expensive, less compatible, and very uncommon.

The Z-CLV recording pattern is easily visible after burning a DVD-R.

Limit

Both DVDs and CDs have been known to explode when damaged and/or spun at excessive speed. This imposes a constraint on the maximum speed (56× for CDs or around 18× in the case of DVDs) at which drives can operate.

An exploded disc.

Loading Mechanisms

Current optical drives use either a *tray-loading mechanism, where the disc is loaded onto a motorized or manually operated tray, or a slot-loading mechanism, where the disc is slid into a slot and drawn in by motorized rollers. With both types of mechanism, if a CD or DVD is left in the drive after the computer is turned off, the disc cannot be ejected using the normal eject mechanism of the drive. However, tray-loading drives account for this situation by providing a small hole where one can insert a straightened paperclip to manually open the drive tray to retrieve the disc. Slot-loading optical disc drives have the disadvantages that they cannot usually accept the smaller 80 mm discs (unless 80 mm optical disc adapter is used) or any non-standard sizes, usually have no emergency eject hole or eject button, and therefore have to be disassembled if the optical disc cannot be ejected normally. However, the Nintendo Wii, because of backwards compatibility with Nintendo GameCube games, and PlayStation 3 video game consoles are able to load standard size DVDs and 80 mm discs in the same slot-loading drive.*

A small number of drive models, mostly compact portable units, have a *top-loading mechanism where the drive lid is opened upwards and the disc is placed directly onto the spindle* (for example, all PlayStation One consoles, portable CD players, and some standalone CD recorders all feature top-loading drives). These sometimes have the advantage of using spring-loaded ball bearings to hold the disc in place, minimizing damage to the disc if the drive is moved while it is spun up.

Some early CD-ROM drives used a mechanism where CDs had to be inserted into special cartridges or caddies, somewhat similar in appearance to a 3.5" floppy diskette. This was intended to protect the disc from accidental damage by enclosing it in a tougher plastic casing, but did not gain wide acceptance due to the additional cost and compatibility concerns—such drives would also inconveniently require "bare" discs to be manually inserted into an openable caddy before use. Ultra Density Optical and Universal Media Disc use optical disc cartridges.

There were also some early CD-ROM drives for desktop PCs in which its tray-loading mechanism will eject slightly and user has to pull out the tray manually to load CD, similar to the tray ejecting method used in internal optical disc drives of modern laptops and modern external slim portable optical disc drives. Like the top-loading mechanism, they have spring-loaded ball bearings on the spindle.

Computer Interfaces

Digital audio output, analog audio output, and parallel ATA interface.

Most internal drives for personal computers, servers and workstations are designed to fit in a standard 5.25" drive bay and connect to their host via an ATA or SATA interface. Additionally, there may be digital and analog outputs for audio. The outputs may be connected via a header cable to the sound card or the motherboard. At one time, computer software resembling cd players controlled playback of the CD. Today the information is extracted from the disc as data, to be played back or converted to other file formats.

External drives usually have USB or FireWire interfaces. Some portable versions for laptops power themselves from batteries or directly from their interface bus.

Drives with SCSI interface were made, but they are less common and tend to be more expensive, because of the cost of their interface chipsets, more complex SCSI connectors, and small volume of sales.

When the optical disc drive was first developed, it was not easy to add to computer systems. Some computers such as the IBM PS/2 were standardizing on the 3.5" floppy and 3.5" hard disk, and did not include a place for a large internal device. Also IBM PCs and clones at first only included a single (parallel) ATA drive interface, which by the time the CDROM was introduced, was already being used to support two hard drives. Early laptops simply had no built-in high-speed interface for supporting an external storage device.

This was solved through several techniques:

- Early sound cards could include a CD-ROM drive interface. Initially, such interfaces were proprietary to each CD-ROM manufacturer. A sound card could often have two or three different interfaces which are able to communicate with cdrom drive.

- A parallel port external drive was developed that connected between a printer and the computer. This was slow but an option for laptops

- A PCMCIA optical drive interface was also developed for laptops

- A SCSI card could be installed in desktop PCs for an external SCSI drive enclosure, though SCSI was typically much more expensive than other options

Internal Mechanism of a Drive

Internal mechanism of a DVD-ROM Drive.

The optical drives in the photos are shown right side up; the disc would sit on top of them. The laser and optical system scans the underside of the disc.

With reference to the top photo, just to the right of image center is the disc motor, a metal cylinder, with a gray centering hub and black rubber drive ring on top. There is a disc-shaped round clamp, loosely held inside the cover and free to rotate; it's not in the photo. After the disc tray stops moving inward, as the motor and its attached parts rise, a magnet near the top of the rotating assembly contacts and strongly attracts the clamp to hold and center the disc. This motor is an "outrunner"-style brushless DC motor which has an external rotor – every visible part of it spins.

Two parallel guide rods that run between upper left and lower right in the photo carry the "sled", the moving optical read-write head. As shown, this "sled" is close to, or at the position where it reads or writes at the edge of the disc. To move the "sled" during continuous read or write operations, a stepper motor rotates a leadscrew to move the "sled" throughout its total travel range. The motor, itself, is the short gray cylinder just to the left of the most-distant shock mount; its shaft is parallel to the support rods. The leadscrew is the rod with evenly-spaced darker details; these are the helical grooves that engage a pin on the "sled".

In contrast, the mechanism shown in the second photo, which comes from a cheaply made DVD player, uses less accurate and less efficient brushed DC motors to both move the sled and spin the disc. Some older drives use a DC motor to move the sled, but also have a magnetic rotary encoder to keep track of the position. Most drives in computers use stepper motors.

The gray metal chassis is shock-mounted at its four corners to reduce sensitivity to external shocks, and to reduce drive noise from residual imbalance when running fast. The soft shock mount grommets are just below the brass-colored screws at the four corners (the left one is obscured).

In the third photo, the components under the cover of the lens mechanism are visible. The two permanent magnets on either side of the lens holder as well as the coils that move the lens can be seen. This allows the lens to be moved up, down, forwards, and backwards to stabilize the focus of the beam.

In the fourth photo, the inside of the optics package can be seen. Note that since this is a CD-ROM drive, there is only one laser, which is the black component mounted to the bottom left of the assembly. Just above the laser are the first focusing lens and prism that direct the beam at the disc. The tall, thin object in the center is a half-silvered mirror that splits the laser beam in multiple directions. To the bottom right of the mirror is the main photodiode that senses the beam reflected off the disc. Above the main photodiode is a second photodiode that is used to sense and regulate the power of the laser.

The irregular orange material is flexible etched copper foil supported by thin sheet plastic; these are "flexible printed circuits" that connect everything to the electronics (which is not shown).

Compatibility

Most optical drives are backwards compatible with their ancestors up to CD, although this is not required by standards.

Compared to a CD's 1.2 mm layer of polycarbonate, a DVD's laser beam only has to penetrate 0.6 mm in order to reach the recording surface. This allows a DVD drive to focus the beam on a smaller spot size and to read smaller pits. DVD lens supports a different focus for CD or DVD media with same laser. With the newer Blu-ray disc drives, the laser only has to penetrate 0.1 mm of material. Thus the optical assembly would normally have to have an even greater focus range. In practice, the Blu-ray optical system is separate from the DVD/CD system.

Optical disc drive	Optical disc or optical media													
	Pressed CD	CD-R	CD-RW	Pressed DVD	DVD-R	DVD +R	DVD- RW	DVD +RW	DVD+R DL	Pressed CAT BD	BD-R	BD- RE	BD-R DL	BD- RE DL
Audio CD player	Read	Read [1]	Read [2]	None	None	None	None	None	None	None	None	None	None	None
CD-ROM drive	Read	Read [1]	Read [2]	None	None	None	None	None	None	None	None	None	None	None
CD-R recorder	Read	Write	Read	None	None	None	None	None	None	None	None	None	None	None
CD-RW recorder	Read	Write	Write	None	None	None	None	None	None	None	None	None	None	None
DVD-ROM drive	Read	Read [3]	Read [3]	Read	Read [4]	Read [4]	Read [4]	Read [4]	Read [5]	None	None	None	None	None
DVD-R recorder	Read	Write	Write	Read	Write	Read [6]	Read	Read [6]	Read [5]	None	None	None	None	None
DVD-RW recorder	Read	Write	Write	Read	Write	Read [7]	Write [8]	Read [6]	Read [5]	None	None	None	None	None
DVD+RW recorder	Read	Write	Write	Read	Read [6]	Read [9]	Read [6]	Write	Read [5]	None	None	None	None	None
DVD+R recorder	Read	Write	Write	Read	Read [6]	Write	Read [6]	Write	Read [5]	None	None	None	None	None
DVD±RW recorder	Read	Write	Write	Read	Write	Write	Write	Write	Read [5]	None	None	None	None	None
DVD±RW/ DVD+R DL recorder	Read	Write	Write	Read	Write [10]	Write	Write [10]	Write	Write	None	None	None	None	None
BD-ROM	Read	Read	Read	Read	Read	Read	Read	Read	Read	Read	Read	Read	Read	Read
BD-R recorder	Read [11]	Write [11]	Write [11]	Read	Write	Write	Write	Write	Write	Read	Write	Read	Read	Read
BD-RE recorder	Read [11]	Write [11]	Write [11]	Read	Write	Write	Write	Write	Write	Read	Write	Write	Read	Read
BD-R DL recorder	Read [11]	Write [11]	Write [11]	Read	Write	Write	Write	Write	Write	Read	Write	Read	Write	Read
BD-RE DL recorder	Read [11]	Write [11]	Write [11]	Read	Write	Write	Write	Write	Write	Read	Write	Write	Write	Write

- Some types of CD-R media with less-reflective dyes may cause problems.

- May not work in non MultiRead-compliant drives.

- May not work in some early-model DVD-ROM drives. CD-R would not work in any drive that did not have a 780 nm laser. CD-RW compatibility varied.

- DVD+RW discs did not work in early video players that played DVD-RW discs. This was not due to any incompatibility with the format but was a deliberate feature built into the firmware by one drive manufacturer.

- Read compatibility with existing DVD drives may vary greatly with the brand of DVD+R DL media used. Also drives that predated the media did not have the book code for DVD+R DL media in their firmware (this was not an issue for DVD-R DL though some drives could only read the first layer).

- Early DVD+RW and DVD+R recorders could not write to DVD-R(W) media (and vice versa).

- Will work in all drives that read DVD-R as compatibility ID byte is the same.

- Recorder firmware may blacklist or otherwise refuse to record to some brands of DVD-RW media.

- DVD+RW format was released before DVD+R. All DVD+RW only drives could be upgraded to write DVD+R discs by a firmware upgrade.

- As of April 2005, all DVD+R DL recorders on the market are Super Multi-capable.

- As of October 2006, recently released BD drives are able to read and write CD media.

Recording Performance

During the times of CD writer drives, they are often marked with three different speed ratings. In these cases, the first speed is for write-once (R) operations, the second speed for re-write (RW) operations, and the last speed for read-only (ROM) operations. For example, a 40×/16×/48× CD writer drive is capable of writing to CD-R media at speed of 40× (6,000 KB/s), writing to CD-RW media at speed of 16× (2,400 KB/s), and reading from a CD-ROM media at speed of 48× (7,200 KB/s).

During the times of combo (CD-RW/DVD-ROM) drives, an additional speed rating (e.g., the 16× in 52×/32×/52×/16×) is designated for DVD-ROM media reading operations.

For DVD writer drives, Blu-ray disc combo drives, and Blu-ray disc writer drives, the writing and reading speed of their respective optical media are specified in its retail box, user's manual, and/or bundled brochures or pamphlets.

In the late 1990s, *buffer underruns became a very common problem as high-speed CD recorders began to appear in home and office computers, which—for a variety of reasons—often could not muster the I/O performance to keep the data stream to the recorder steadily fed. The recorder, should it run short, would be forced to halt the recording process, leaving a truncated track that usually renders the disc useless.*

In response, manufacturers of CD recorders began shipping drives with "buffer underrun protection" (under various trade names, such as Sanyo's "BURN-Proof", Ricoh's "JustLink" and Yamaha's "Lossless Link"). These can suspend and resume the recording process in such a way that the gap the stoppage produces can be dealt with by the error-correcting logic built into CD players and CD-ROM drives. The first of these drives were rated at 12× and 16×.

While drives are burning DVD+R, DVD+RW and all Blu-ray formats, they do not require any such error correcting recovery as the recorder is able to place the new data exactly on the end of the suspended write effectively producing a continuous track (this is what the DVD+ technology achieved). Although later interfaces were able to stream data at the required speed, many drives now write in a 'zoned constant linear velocity'. This means that the drive has to temporarily suspend the write operation while it changes speed and then recommence it once the new speed is attained. This is handled in the same manner as a buffer underrun.

The internal buffer of optical disc writer drives is: 8 MiB when recording BD-R/BD-R DL/BD-RE/BD-RE DL media; 2 MiB when recording DVD-R/DVD-RW/DVD-R DL/DVD+R/DVD+RW/DVD+RW DL/DVD-RAM/CD-R/CD-RW media.

Recording Schemes

CD recording on personal computers was originally a batch-oriented task in that it required specialised authoring software to create an "image" of the data to record, and to record it to disc in the one session. This was acceptable for archival purposes, but limited the general convenience of CD-R and CD-RW discs as a removable storage medium.

Packet writing is a scheme in which the recorder writes incrementally to disc in short bursts, or packets. Sequential packet writing fills the disc with packets from bottom up. To make it readable in CD-ROM and DVD-ROM drives, the disc can be *closed at any time by writing a final table-of-contents to the start of the disc; thereafter, the disc cannot be packet-written any further. Packet writing, together with support from the* operating system and a file system like UDF, can be used to mimic random write-access as in media like flash memory and magnetic disks.

Fixed-length packet writing (on CD-RW and DVD-RW media) divides up the disc into padded, fixed-size packets. The padding reduces the capacity of the disc, but allows the recorder to start and stop recording on an individual packet without affecting its neighbours. These resemble the block-writable access offered by magnetic media closely enough that many conventional file systems will work as-is. Such discs, however, are not readable in most CD-ROM and DVD-ROM drives or on most operating systems without additional third-party drivers. The division into packets is not as reliable as it may seem as CD-R(W) and DVD-R(W) drives can only locate data to within a data block. Although generous gaps (the padding referred to above) are left between blocks, the drive nevertheless can occasionally miss and either destroy some existing data or even render the disc unreadable.

The DVD+RW disc format eliminates this unreliability by embedding more accurate timing hints in the data groove of the disc and allowing individual data blocks (or even bytes) to be replaced without affecting backwards compatibility (a feature dubbed "lossless linking"). The format itself was designed to deal with discontinuous recording because it was expected to be widely used in digital video recorders. Many such DVRs use variable-rate video compression schemes which require them to record in short bursts; some allow simultaneous playback and recording by alternating quickly between recording to the tail of the disc whilst reading from elsewhere. The Blu-ray disc system also encompasses this technology.

Mount Rainier aims to make packet-written CD-RW and DVD+RW discs as convenient to use as that of removable magnetic media by having the firmware format new discs in the background and manage media defects (by automatically mapping parts of the disc which have been worn out by erase cycles to reserve space elsewhere on the disc). As of February 2007, support for Mount Rainier is natively supported in Windows Vista. All previous versions of Windows require a third-party solution, as does Mac OS X.

Recorder Unique Identifier

Owing to pressure from the music industry, as represented by the IFPI and RIAA, Philips developed the *Recorder Identification Code (RID) to allow media to be uniquely associated with the recorder that has written it. This standard is contained in the* Rainbow Books. The RID-Code consists of a supplier code (e.g. "PHI" for Philips), a model number and the unique ID of the recorder. Quoting Philips, the RID "enables a trace for each disc back to the exact machine on which it was made using coded information in the recording itself. The use of the RID code is mandatory."

Although the RID was introduced for music and video industry purposes, the RID is included on every disc written by every drive, including data and backup discs. The value of the RID is questionable as it is (currently) impossible to locate any individual recorder due to there being no database.

Source IDentification Code

The Source IDentification Code (SID) is an eight character supplier code that is placed on optical discs by the manufacturer. The SID identifies not only manufacturer, but also the individual factory and machine that produced the disc.

According to Phillips, the administrator of the SID codes, the SID code provides an optical disc production facility with the means to identify all discs mastered and/or replicated in its plant, including the specific Laser Beam Recorder (LBR) signal processor or mould that produced a particular stamper or disc.

Use of RID and SID together in forensics

The standard use of RID and SID mean that each disc written contains a record of the machine that produced a disc (the SID), and which drive wrote it (the RID). This combined knowledge may be very useful to law enforcement, to investigative agencies, and to private and/or corporate investigators.

Laser Printing

HP LaserJet 4200 series printer, installed atop high-capacity paper feeder

Laser printing is an electrostatic digital printing process. It produces high-quality text and graphics (and moderate-quality photographs) by repeatedly passing a laser beam back and forth over a negatively charged cylinder called a "drum" to define a differentially charged image. The drum then selectively collects electrically charged powdered ink (toner), and transfers the image to paper, which is then heated in order to permanently fuse the text and/or imagery. As with digital photocopiers and multifunction/all-in-one inkjet printers, laser printers employ a xerographic printing process. However, laser printing differs from analog photocopiers in that the image is produced by the direct scanning of the medium across the printer's photoreceptor. This enables laser printing to copy images more quickly than most photocopiers.

Invented at Xerox PARC in the 1970s, laser printers were introduced for the office and then home markets in subsequent years by IBM, Canon, Xerox, Apple, Hewlett-Packard and many others. Over the decades, quality and speed have increased as price has fallen, and the once cutting-edge printing devices are now ubiquitous.

History

Gary Starkweather invented the laser printer (2009 photo)

In the 1960s, the Xerox Corporation held a dominant position in the photocopier market. In 1969, Gary Starkweather, who worked in Xerox's product development department, had the idea of using a laser beam to "draw" an image of what was to be copied directly onto the copier drum. After transferring to the recently formed Palo Alto Research Center (Xerox PARC) in 1971, Starkweather adapted a Xerox 7000 copier to create SLOT (Scanned Laser Output Terminal). In 1972, Starkweather worked with Butler Lampson and Ronald Rider to add a control system and character generator, resulting in a printer called EARS (Ethernet, Alto Research character generator, Scanned laser output terminal) -- which later became the Xerox 9700 laser printer.

The first commercial implementation of a laser printer was the IBM 3800 in 1976. It was designed for data centers, where it replaced line printers attached to mainframe computers. The IBM 3800 was used for high-volume printing on continuous stationery, and achieved speeds of 215 pages per minute (ppm), at a resolution of 240 dots per inch (dpi). Over 8,000 of these printers were sold. The Xerox 9700 was brought to market in 1977. Unlike the IBM 3800, the Xerox 9700 was not targeted to replace any particular existing printers; but, it did have limited support for the loading of fonts. The Xerox 9700 excelled at printing high-value documents on cut-sheet paper with varying content (e.g. insurance policies).

In 1979, inspired by the Xerox 9700's commercial success, Japanese camera and optics company, Canon, developed a low-cost, *desktop laser printer: the Canon LBP-10. Canon then began work on a much-improved print engine, the Canon CX, resulting in the LBP-CX printer. Lacking experience in selling to computer users, Canon sought partnerships with three* Silicon Valley companies: Diablo Data Systems (who turned them down), Hewlett-Packard (HP), and Apple Computer.

The first laser printer designed for office use reached market in 1981: the Xerox Star 8010. The system used a desktop metaphor that was unsurpassed in commercial sales, until the Apple Macintosh. Although it was innovative, the Star workstation was a prohibitively expensive (US$17,000) system, affordable only to a fraction of the businesses and institutions at which it was targeted.

The first laser printer intended for mass-market sales was the HP LaserJet, released in 1984; it used the Canon CX engine, controlled by HP software. The LaserJet was quickly followed by printers from Brother Industries, IBM, and others. First-generation machines had large photosensitive drums, of circumference greater than the loaded paper's length. Once faster-recovery coatings were developed, the drums could touch the paper multiple times in a pass, and therefore be smaller in diameter.

In 1985, Apple introduced the LaserWriter (also based on the Canon CX engine), but used the newly released PostScript page-description language. Up until this point, each manufacturer used its own proprietary page-description language, making the supporting software complex and expensive. PostScript allowed the use of text, fonts, graphics, images, and color largely independent of the printer's brand or resolution. PageMaker, written by Aldus for the Macintosh and LaserWriter, was also released in 1985 and the combination became very popular for desktop publishing.:13/23:364 Laser printers brought exceptionally fast and high-quality text printing in multiple fonts on a page, to the business and consumer markets. No other commonly available printer during this era could also offer this combination of features.

Printing Process

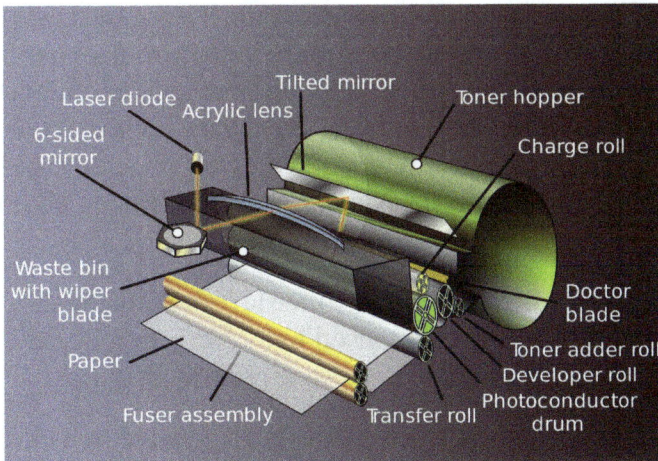

Diagram of a laser printer

A laser beam (typically, an aluminium gallium arsenide (AlGaAs) semiconductor laser) projects an image of the page to be printed onto an electrically charged, selenium-coated, rotating, cylindrical drum (or, more commonly in subsequent versions, a drum called an organic photoconductor made of N-vinylcarbazole, an organic monomer). Photoconductivity allows the charged electrons to fall away from the areas exposed to light. Powdered ink (toner) particles are then electrostatically attracted to the charged areas of the drum that have not been laser-beamed. The drum then transfers the image onto paper (which is passed through the machine) by direct contact. Finally the paper is passed onto a finisher, which uses intense heat to instantly fuse the toner/image onto the paper.

There are typically seven steps involved in the process:

Raster Image Processing

The document to be printed is encoded in a page description language such as PostScript, Printer Command Language (PCL), or Open XML Paper Specification (OpenXPS). The raster image processor converts the page description into a bitmap which is stored in the printer's raster memory. Each horizontal strip of dots across the page is known as a raster line or scan line.

Laser printing differs from other printing technologies in that each page is always rendered in a single continuous process without any pausing in the middle, while other technologies like inkjet can pause every few lines. To avoid a buffer underrun (where the laser reaches a point on the page before it has the dots to draw there), a laser printer typically needs enough raster memory to hold the bitmap image of an entire page.

Memory requirements increase with the square of the dots per inch, so 600 dpi requires a minimum of 4 megabytes for monochrome, and 16 megabytes for color at 600 dpi. For fully graphical output using a page description language, a minimum of 1 megabyte of memory is needed to store an entire monochrome letter/A4 sized page of dots at 300 dpi. At 300 dpi, there are 90,000 dots per square inch (300 dots per linear inch). A typical 8.5 × 11 sheet of paper has 0.25-inch (6.4 mm) margins, reducing the printable area to 8.0 by 10.5 inches (200 mm × 270 mm), or 84 square inches. 84 sq/in × 90,000 dots per sq/in = 7,560,000 dots. 1 megabyte = 1,048,576 bytes, or 8,388,608 bits, which is just large enough to hold the entire page at 300 dpi, leaving about 100 kilobytes to spare for use by the raster image processor.

In a color printer, each of the four CMYK toner layers is stored as a separate bitmap, and all four layers are typically preprocessed before printing begins, so a minimum of 4 megabytes is needed for a full-color letter-size page at 300 dpi.

During the 1980s, memory chips were still very expensive, which is why entry-level laser printers in that era always came with four-digit suggested retail prices in US dollars. Memory prices later plunged, and 1200 dpi printers have been widely available in the consumer market since 2008. 2400 dpi electrophotographic printing plate makers, essentially laser printers that print on plastic sheets, are also available.

Charging

In older printers, a corona wire positioned parallel to the drum or, in more recent printers, a primary charge roller, projects an electrostatic charge onto the photoreceptor (otherwise named the photo conductor unit), a revolving photosensitive drum or belt, which is capable of holding an electrostatic charge on its surface while it is in the dark.

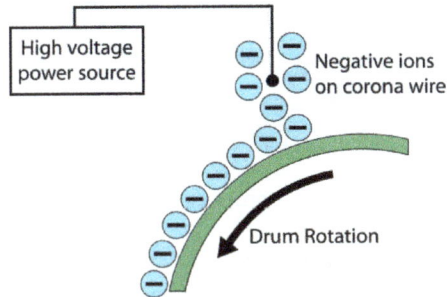

Applying a negative charge to the photosensitive drum

An AC bias voltage is applied to the primary charge roller to remove any residual charges left by previous images. The roller will also apply a DC bias on the drum surface to ensure a uniform negative potential.

Numerous patents[specify] describe the photosensitive drum coating as a silicon sandwich with a photocharging layer, a charge leakage barrier layer, as well as a surface layer. One version[specify] uses amorphous silicon containing hydrogen as the light receiving layer, Boron nitride as a charge leakage barrier layer, as well as a surface layer of doped silicon, notably silicon with oxygen or nitrogen which at sufficient concentration resembles machining silicon nitride.

Exposing

A laser printer uses a laser because lasers are able to form highly focused, precise, and intense beams of light, especially over the short distances inside of a printer. The laser is aimed at a rotating polygonal mirror which directs the light beam through a system of lenses and mirrors onto the photoreceptor drum, writing pixels at rates up to sixty five million times per second. The drum continues to rotate during the sweep, and the angle of sweep is canted very slightly to compensate for this motion. The stream of rasterized data held in the printer's memory rapidly turns the laser on and off as it sweeps.

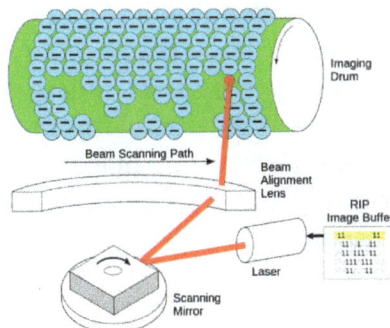

Laser light selectively neutralizes the negative charge on the photoreceptive drum, to form an electrostatic image

Laser unit from a Dell P1500. The white hexagon is the rotating scanner mirror.

The laser beam neutralizes (or reverses) the charge on the surface of the drum, leaving a static electric negative image on the drum's surface which will repel the negatively charged toner particles. The areas on the drum which were struck by the laser, however, momentarily have no charge, and the toner being pressed against the drum by the toner-coated developer roll in the next step moves from the roll's rubber surface to the uncharged portions of the surface of the drum.

Some non-laser printers (LED printers) use an array of light emitting diodes spanning the width of the page to generate an image, rather than using a laser. "Exposing" is also known as "writing" in some documentation.

Developing

The surface with the latent image is exposed to toner which has been applied in a 15-micron-thick layer to the developer roll. Toner consists of fine particles of dry plastic powder mixed with carbon black or coloring agents. The toner particles are given a negative charge inside the toner cartridge, and as they emerge onto the developer drum they are electrostatically attracted to the photoreceptor's latent image (the areas on the surface of the drum which had been struck by the laser). Because negative charges repel each other, the negatively charged toner particles will not adhere to the drum where the negative charge (imparted previously by the charge roller) remains.

Transferring

A sheet of paper is then rolled under the photoreceptor drum, which has been coated with a pattern of toner particles in the exact places where the laser struck it moments before. The toner particles have a very weak attraction to both the drum and the paper, but the bond to the drum is weaker and the particles transfer once again, this time from the drum's surface to the paper's surface. Some machines also use a positively charged "transfer roller" on the back side of the paper to help pull the negatively charged toner from the photoreceptor drum to the paper.

Fusing

The paper passes through rollers in the fuser assembly, where temperatures up to

200 °C (392 °F) and pressure are used to permanently bond the plastic powder to the paper. One roller is usually a hollow tube (heat roller) and the other is a rubber backed roller (pressure roller). A radiant heat lamp is suspended in the centre of the hollow tube, and its infrared energy uniformly heats the roller from the inside. For proper bonding of the toner, the fuser roller must be uniformly hot.

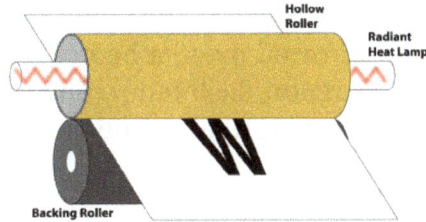

Toner is melted onto paper with heat and pressure

Some printers use a very thin flexible metal foil roller, so there is less thermal mass to be heated and the fuser can more quickly reach operating temperature. If paper moves through the fuser more slowly, there is more roller contact time for the toner to melt, and the fuser can operate at a lower temperature. Smaller, inexpensive laser printers typically print slowly, due to this energy-saving design, compared to large high speed printers where paper moves more rapidly through a high-temperature fuser with a very short contact time.

Cleaning and Recharging

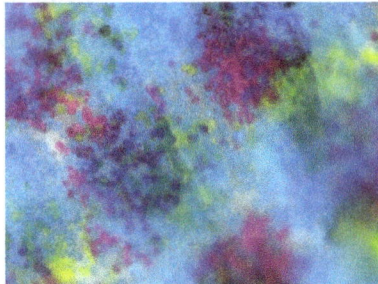

Magnification of color laser printer output, showing individual toner particles comprising 4 dots of an image with a bluish background

As the drum completes a revolution, it is exposed to an electrically neutral soft plastic blade which cleans any remaining toner from the photoreceptor drum and deposits it into a waste reservoir. A charge roller then re-establishes a uniform negative charge on the surface of the now clean drum, readying it to be struck again by the laser.

Continuous Printing

Once the raster image generation is complete, all steps of the printing process can occur one after the other in rapid succession. This permits the use of a very small and compact unit, where the photoreceptor is charged, rotates a few degrees and is scanned, rotates a few more degrees and is developed, and so forth. The entire process can be completed before the drum completes one revolution.

Different printers implement these steps in distinct ways. LED printers actually use a linear array of light-emitting diodes to "write" the light on the drum. The toner is based on either wax or plastic, so that when the paper passes through the fuser assembly, the particles of toner melt. The paper may or may not be oppositely charged. The fuser can be an infrared oven, a heated pressure roller, or (on some very fast, expensive printers) a xenon flash lamp. The warmup process that a laser printer goes through when power is initially applied to the printer consists mainly of heating the fuser element.

Malfunctions

The mechanism inside a laser printer is somewhat delicate and, once damaged, often impossible to repair. The drum in particular is a critical component: it must not be left exposed to ambient light for more than a few hours, as light is what causes it to lose its charge and will eventually wear it out. Anything that interferes with the operation of the laser such as a scrap of torn paper may prevent the laser from discharging some portion of the drum, causing those areas to appear as white vertical streaks. If the neutral wiper blade fails to remove residual toner from the drum's surface, that toner may circulate on the drum a second time, causing smears on the printed page with each revolution. If the charge roller becomes damaged or does not have enough power, it may fail to adequately negatively charge the surface of the drum, allowing the drum to pick up excessive toner on the next revolution from the developer roll and causing a repeated but fainter image from the previous revolution to appear down the page.

If the toner doctor blade does not ensure that a smooth, even layer of toner is applied to the developer roll, the resulting printout may have white streaks from this in places where the blade has scraped off too much toner. Alternatively if the blade allows too much toner to remain on the developer roll, the toner particles might come loose as the roll turns, precipitate onto the paper below, and become bonded to the paper during the fusing process. This will result in a general darkening of the printed page in broad vertical stripes with very soft edges.

If the fuser roller does not reach a high enough temperature or if the ambient humidity is too high, the toner will not fuse well to the paper and may flake off after printing. If the fuser is too hot, the plastic component of the toner may smear, causing the printed text to look like it is wet or smudged, or may cause the melted toner to soak through the paper to the back side.

Different manufacturers claim that their toners are specifically developed for their printers, and that other toner formulations may not match the original specifications in terms of either tendency to accept a negative charge, to move to the discharged areas of the photoreceptor drum from the developer roll, to fuse appropriately to the paper, or to come off the drum cleanly in each revolution.

Performance

As with most electronic devices, the cost of laser printers has fallen markedly over the years. In 1984, the HP LaserJet sold for $3500, had trouble with even small, low resolution graphics, and weighed 32 kg (71 lb). As of 2016[update], low-end monochrome laser printers can sell for less than $75. These printers tend to lack onboard processing and rely on the host computer to generate a raster image, but outperform the 1984 LaserJet in nearly all situations.

Laser printer speed can vary widely, and depends on many factors, including the graphic intensity of the job being processed. The fastest models can print over 200 monochrome pages per minute (12,000 pages per hour). The fastest color laser printers can print over 100 pages per minute (6000 pages per hour). Very high-speed laser printers are used for mass mailings of personalized documents, such as credit card or utility bills, and are competing with lithography in some commercial applications.

The cost of this technology depends on a combination of factors, including the cost of paper, toner, drum replacement, as well as the replacement of other items such as the fuser assembly and transfer assembly. Often printers with soft plastic drums can have a very high cost of ownership that does not become apparent until the drum requires replacement.

Duplex printing (printing on both sides of the paper) can halve paper costs and reduce filing volumes. Formerly only available on high-end printers, duplexers are now common on mid-range office printers, though not all printers can accommodate a duplexing unit. Duplexing can also give a slower page-printing speed, because of the longer paper path.

Color Laser Printers

Fuji Xerox color laser printer C1110B

Color laser printers use colored toner (dry ink), typically cyan, magenta, yellow, and black (CMYK). While monochrome printers only use one laser scanner assembly, color printers often have two or more.

Color printing adds complexity to the printing process because very slight misalignments known as registration errors can occur between printing each color, causing unintended color fringing, blurring, or light/dark streaking along the edges of colored regions. To permit a high registration accuracy, some color laser printers use a large rotating belt called a "transfer belt". The transfer belt passes in front of all the toner cartridges and each of the toner layers are precisely applied to the belt. The combined layers are then applied to the paper in a uniform single step.

Color printers usually have a higher cost per page than monochrome printers (even if printing monochrome-only pages).

Business Model Comparison with Inkjet Printers

Manufacturers use a similar business model for both low-cost color laser printers and inkjet printers: the printers are sold cheaply while replacement toners and inks are relatively expensive. Color laser printers are much faster than inkjet printers and their running cost per page is usually slightly less. The print quality of color lasers is limited by their resolution (typically 600–1200 dpi) and their use of just four color toners. They often have trouble printing large areas of the same or subtle gradations of color. Inkjet printers designed for printing photos can produce much higher quality color images.

Anti-Counterfeiting Marks

Small yellow dots on white paper, generated by a color laser printer, are nearly invisible.

Many modern color laser printers mark printouts by a nearly invisible dot raster, for the purpose of traceability. The dots are yellow and about 0.1 mm (0.0039 in) in size, with a raster of about 1 mm (0.039 in). This is purportedly the result of a deal between the US government and printer manufacturers to help track counterfeiters. The dots encode data such as printing date, time, and printer serial number in binary-coded decimal on every sheet of paper printed, which allows pieces of paper to be traced by the manufacturer to identify the place of purchase, and sometimes the buyer.

Digital rights advocacy groups such as the Electronic Frontier Foundation are concerned about this erosion of the privacy and anonymity of those who print.

Smart Chips in Toner Cartridges

Similar to inkjet printers, toner cartridges may contain smart chips that reduce the number of pages that can be printed with it (reducing the amount of usable ink in the cartridge to sometimes only 50%), in an effort to increase sales of the toner cartridges. Besides being more expensive to the consumer, this technique also increases waste, and thus increases pressure on the environment. For these toner cartridges (as with inkjet cartridges), reset devices can be used to override the limitation set by the smart chip. Also, for some printers, online walk-throughs have been posted to demonstrate how to use up all the ink in the cartridge. These chips offer no benefit to the end consumer— all laser printers originally used an optical mechanism to assess the amount of remaining toner in the cartridge rather than using a chip to electrically count the number of printed pages, and the chip's only function was as an alternate method to decrease the cartridge's usable life.

Safety Hazards, Health Risks, and Precautions

Toner Clean-Up

Toner particles are formulated to have electrostatic properties and can develop static electric charges when they rub against other particles, objects, or the interiors of transport systems and vacuum hoses. Static discharge from charged toner particles can ignite combustible particles in a vacuum cleaner bag or create a small dust explosion if sufficient toner is airborne. Toner particles are so fine that they are poorly filtered by conventional household vacuum cleaner filter bags and blow through the motor or back into the room.

If toner spills into the laser printer, a special type of vacuum cleaner with an electrically conductive hose and a high-efficiency (HEPA) filter may be needed for effective cleaning. These specialized tools are called "ESD-safe" (Electrostatic Discharge-safe) or "toner vacuums".

Ozone Hazards

As a normal part of the printing process, the high voltages inside the printer can produce a corona discharge that generates a small amount of ionized oxygen and nitrogen, which react to form ozone and nitrogen oxides. In larger commercial printers and copiers, an activated carbon filter in the air exhaust stream breaks down these noxious gases to prevent pollution of the office environment.

However, some ozone escapes the filtering process in commercial printers, and ozone filters are not used at all in most smaller consumer printers. When a laser printer or copier is operated for a long period of time in a small, poorly ventilated space, these gases can build up to levels at which the odor of ozone or irritation may be noticed. A potential for creating a health hazard is theoretically possible in extreme cases.

Respiratory Health Risks

According to a 2012 study conducted in Queensland, Australia, some printers emit sub-micrometre particles which some suspect may be associated with respiratory diseases. Of 63 printers evaluated in the Queensland University of Technology study, 17 of the strongest emitters were made by HP and one by Toshiba. The machine population studied, however, was only those machines already in place in the building and was thus biased toward specific manufacturers. The authors noted that particle emissions varied substantially even among the same model of machine. According to Professor Morawska of Queensland University, one printer emitted as many particles as a burning cigarette:

The health effects from inhaling ultrafine particles depend on particle composition, but the results can range from respiratory irritation to more severe illness such as cardiovascular problems or cancer.

— *Queensland University of Technology*

Muhle et al. (1991) reported that the responses to chronically inhaled copying toner, a plastic dust pigmented with carbon black, titanium dioxide and silica were also similar qualitatively to titanium dioxide and diesel exhaust.

In December 2011, the Australian government agency Safe Work Australia reviewed existing research and concluded that "no epidemiology studies directly associating laser printer emissions with adverse health outcomes were located" and that several assessments conclude that "risk of direct toxicity and health effects from exposure to laser printer emissions is negligible". The review also observes that, because the emissions have been shown to be volatile or semi-volatile organic compounds, "it would be logical to expect possible health effects to be more related to the chemical nature of the aerosol rather than the physical character of the 'particulate' since such emissions are unlikely to be or remain as 'particulates' after they come into contact with respiratory tissue".

Air Transport Ban

After the 2010 cargo plane bomb plot, in which shipments of laser printers with explosive-filled toner cartridges were discovered on separate cargo airplanes, the US Transportation Security Administration prohibited pass-through passengers from carrying toner or ink cartridges weighing over 1 pound (0.45 kg) on inbound flights, in both carry-on and checked luggage. *PC Magazine noted that the ban would not impact most travelers, as the majority of cartridges do not exceed the proscribed weight.*

3D scanner

A 3D scanner is a device that analyses a real-world object or environment to collect data on its shape and possibly its appearance (e.g. colour). The collected data can then be used to construct digital three-dimensional models.

Three-dimensional scanner used to create 3D animation and special effects.

Many different technologies can be used to build these 3D-scanning devices; each technology comes with its own limitations, advantages and costs. Many limitations in the kind of objects that can be digitised are still present, for example, optical technologies encounter many difficulties with shiny, mirroring or transparent objects. For example, industrial computed tomography scanning can be used to construct digital 3D models, applying non-destructive testing.

3D scanned interior of St Joseph's Church, Subiaco.

Collected 3D data is useful for a wide variety of applications. These devices are used extensively by the entertainment industry in the production of movies and video games. Other common applications of this technology include industrial design, orthotics and prosthetics, reverse engineering and prototyping, quality control/inspection and documentation of cultural artifacts.t

Functionality

The purpose of a 3D scanner is usually to create a point cloud of geometric samples on the surface of the subject. These points can then be used to extrapolate the shape of the subject (a process called reconstruction). If colour information is collected at each point, then the colours on the surface of the subject can also be determined.

3D scanning of a fin whale skeleton in the Natural History Museum of Slovenia (August 2013)

3D scanners share several traits with cameras. Like most cameras, they have a cone-like field of view, and like cameras, they can only collect information about surfaces that are not obscured. While a camera collects colour information about surfaces within its field of view, a 3D scanner collects distance information about surfaces within its field of view. The "picture" produced by a 3D scanner describes the distance to a surface at each point in the picture. This allows the three dimensional position of each point in the picture to be identified.

For most situations, a single scan will not produce a complete model of the subject. Multiple scans, even hundreds, from many different directions are usually required to obtain information about all sides of the subject. These scans have to be brought into a common reference system, a process that is usually called *alignment or registration, and then merged to create a complete model. This whole process, going from the single range map to the whole model, is usually known as the 3D scanning pipeline.*

Technology

There are a variety of technologies for digitally acquiring the shape of a 3D object. A well established classification divides them into two types: contact and non-contact. Non-contact solutions can be further divided into two main categories, active and passive. There are a variety of technologies that fall under each of these categories.

Contact

Contact 3D scanners probe the subject through physical touch, while the object is in contact with or resting on a precision flat surface plate, ground and polished to a specific maximum of surface roughness. Where the object to be scanned is not flat or can not rest stably on a flat surface, it is supported and held firmly in place by a fixture.

The scanner mechanism may have three different forms:

A carriage system with rigid arms held tightly in perpendicular relationship and each axis gliding along a track. Such systems work best with flat profile shapes or simple convex curved surfaces.

A coordinate measuring machine with rigid perpendicular arms.

An articulated arm with rigid bones and high precision angular sensors. The location of the end of the arm involves complex math calculating the wrist rotation angle and hinge angle of each joint. This is ideal for probing into crevasses and interior spaces with a small mouth opening.

A combination of both methods may be used, such as an articulated arm suspended from a traveling carriage, for mapping large objects with interior cavities or overlapping surfaces.

A CMM (coordinate measuring machine) is an example of a contact 3D scanner. It is used mostly in manufacturing and can be very precise. The disadvantage of CMMs though, is that it requires contact with the object being scanned. Thus, the act of scanning the object might modify or damage it. This fact is very significant when scanning delicate or valuable objects such as historical artifacts. The other disadvantage of CMMs is that they are relatively slow compared to the other scanning methods. Physically moving the arm that the probe is mounted on can be very slow and the fastest CMMs can only operate on a few hundred hertz. In contrast, an optical system like a laser scanner can operate from 10 to 500 kHz.

Other examples are the hand driven touch probes used to digitise clay models in computer animation industry.

Non-Contact Active

Active scanners emit some kind of radiation or light and detect its reflection or radiation passing through object in order to probe an object or environment. Possible types of emissions used include light, ultrasound or x-ray.

Time-Of-Flight

The time-of-flight 3D laser scanner is an active scanner that uses laser light to probe the subject. At the heart of this type of scanner is a time-of-flight laser range finder.

The laser range finder finds the distance of a surface by timing the round-trip time of a pulse of light. A laser is used to emit a pulse of light and the amount of time before the reflected light is seen by a detector is measured. Since the speed of light c is known, the round-trip time determines the travel distance of the light, which is twice the distance between the scanner and the surface. If t is the round-trip time, then distance is equal to $c \cdot t / 2$. The accuracy of a time-of-flight 3D laser scanner depends on how precisely we can measure the t time: 3.3 picoseconds (approx.) is the time taken for light to travel 1 millimetre.

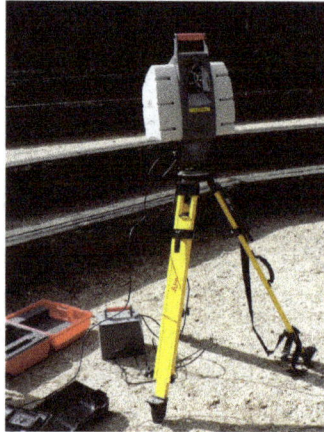

This lidar scanner may be used to scan buildings, rock formations, etc., to produce a 3D model. The lidar can aim its laser beam in a wide range: its head rotates horizontally, a mirror flips vertically. The laser beam is used to measure the distance to the first object on its path.

The laser range finder only detects the distance of one point in its direction of view. Thus, the scanner scans its entire field of view one point at a time by changing the range finder's direction of view to scan different points. The view direction of the laser range finder can be changed either by rotating the range finder itself, or by using a system of rotating mirrors. The latter method is commonly used because mirrors are much lighter and can thus be rotated much faster and with greater accuracy. Typical time-of-flight 3D laser scanners can measure the distance of 10,000~100,000 points every second.

Time-of-flight devices are also available in a 2D configuration. This is referred to as a time-of-flight camera.

Triangulation

Triangulation based 3D laser scanners are also active scanners that use laser light to probe the environment. With respect to time-of-flight 3D laser scanner the triangulation laser shines a laser on the subject and exploits a camera to look for the location of the laser dot. Depending on how far away the laser strikes a surface, the laser dot appears at different places in the camera's field of view. This technique is called triangulation because the laser dot, the camera and the laser emitter form a triangle. The length of one side of the triangle, the distance between the camera and the laser emitter is known. The angle of the laser emitter corner is also known. The angle of the camera corner can be determined by looking at the location of the laser dot in the camera's field of view. These three pieces of information fully determine the shape and size of the triangle and give the location of the laser dot corner of the triangle. In most cases a laser stripe, instead of a single laser dot, is swept across the object to speed up the acquisition process. The National Research Council of Canada was among the first institutes to develop the triangulation based laser scanning technology in 1978.

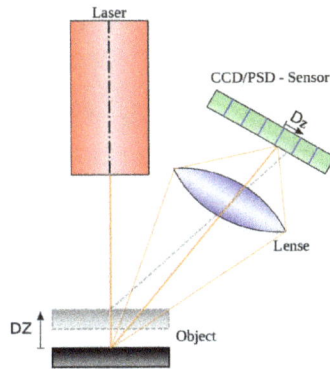

Principle of a laser triangulation sensor. Two object positions are shown.

Strengths and Weaknesses

Time-of-flight and triangulation range finders each have strengths and weaknesses that make them suitable for different situations. The advantage of time-of-flight range finders is that they are capable of operating over very long distances, on the order of kilometres. These scanners are thus suitable for scanning large structures like buildings or geographic features. The disadvantage of time-of-flight range finders is their accuracy. Due to the high speed of light, timing the round-trip time is difficult and the accuracy of the distance measurement is relatively low, on the order of millimetres.

Triangulation range finders are exactly the opposite. They have a limited range of some meters, but their accuracy is relatively high. The accuracy of triangulation range finders is on the order of tens of micrometers.

Time-of-flight scanners' accuracy can be lost when the laser hits the edge of an object because the information that is sent back to the scanner is from two different locations for one laser pulse. The coordinate relative to the scanner's position for a point that has hit the edge of an object will be calculated based on an average and therefore will put the point in the wrong place. When using a high resolution scan on an object the chances of the beam hitting an edge are increased and the resulting data will show noise just behind the edges of the object. Scanners with a smaller beam width will help to solve this problem but will be limited by range as the beam width will increase over distance. Software can also help by determining that the first object to be hit by the laser beam should cancel out the second.

At a rate of 10,000 sample points per second, low resolution scans can take less than a second, but high resolution scans, requiring millions of samples, can take minutes for some time-of-flight scanners. The problem this creates is distortion from motion. Since each point is sampled at a different time, any motion in the subject or the scanner will distort the collected data. Thus, it is usually necessary to mount both the subject and the scanner on stable platforms and minimise vibration. Using these scanners to scan objects in motion is very difficult.

Recently, there has been research on compensating for distortion from small amounts of vibration and distortions due to motion and/or rotation.

When scanning in one position for any length of time slight movement can occur in the scanner position due to changes in temperature. If the scanner is set on a tripod and there is strong sunlight on one side of the scanner then that side of the tripod will expand and slowly distort the scan data from one side to another. Some laser scanners have level compensators built into them to counteract any movement of the scanner during the scan process.

chutyapa

In a conoscopic system, a laser beam is projected onto the surface and then the immediate reflection along the same ray-path are put through a conoscopic crystal and projected onto a CCD. The result is a diffraction pattern, that can be frequency analyzed to determine the distance to the measured surface. The main advantage with conoscopic holography is that only a single ray-path is needed for measuring, thus giving an opportunity to measure for instance the depth of a finely drilled hole.

Using a periscope allows to into small diameter holes and measure bottom and side walls.

Hand-Held Laser Scanners

Hand-held laser scanners create a 3D image through the triangulation mechanism described above: a laser dot or line is projected onto an object from a hand-held device and a sensor (typically a charge-coupled device or position sensitive device) measures the distance to the surface. Data is collected in relation to an internal coordinate system and therefore to collect data where the scanner is in motion the position of the scanner must be determined. The position can be determined by the scanner using reference features on the surface being scanned (typically adhesive reflective tabs, but natural features have been also used in research work) or by using an external tracking method. External tracking often takes the form of a laser tracker (to provide the sensor position) with integrated camera (to determine the orientation of the scanner) or a photogrammetric solution using 3 or more cameras providing the complete Six degrees of freedom of the scanner. Both techniques tend to use infra red Light-emitting diodes attached to the scanner which are seen by the camera(s) through filters providing resilience to ambient lighting.

Data is collected by a computer and recorded as data points within Three-dimensional space, with processing this can be converted into a triangulated mesh and then a Computer-aided design model, often as Non-uniform rational B-spline surfaces. Hand-held laser scanners can combine this data with passive, visible-light sensors — which capture surface textures and colours — to build (or "reverse engineer") a full 3D model.

Structured Light

Structured-light 3D scanners project a pattern of light on the subject and look at the deformation of the pattern on the subject. The pattern is projected onto the subject using either an LCD projector or other stable light source. A camera, offset slightly from the pattern projector, looks at the shape of the pattern and calculates the distance of every point in the field of view.

Structured-light scanning is still a very active area of research with many research papers published each year. Perfect maps have also been proven useful as structured light patterns that solve the correspondence problem and allow for error detection and error correction.

The advantage of structured-light 3D scanners is speed and precision. Instead of scanning one point at a time, structured light scanners scan multiple points or the entire field of view at once. Scanning an entire field of view in a fraction of a second reduces or eliminates the problem of distortion from motion. Some existing systems are capable of scanning moving objects in real-time. VisionMaster creates a 3D scanning system with a 5-megapixel camera – 5 million data points are acquired in every frame.

A real-time scanner using digital fringe projection and phase-shifting technique (certain kinds of structured light methods) was developed, to capture, reconstruct, and render high-density details of dynamically deformable objects (such as facial expressions) at 40 frames per second. Recently, another scanner has been developed. Different patterns can be applied to this system, and the frame rate for capturing and data processing achieves 120 frames per second. It can also scan isolated surfaces, for example two moving hands. By utilising the binary defocusing technique, speed breakthroughs have been made that could reach hundreds of to thousands of frames per second.

Modulated Light

Modulated light 3D scanners shine a continually changing light at the subject. Usually the light source simply cycles its amplitude in a sinusoidal pattern. A camera detects the reflected light and the amount the pattern is shifted by determines the distance the light travelled. Modulated light also allows the scanner to ignore light from sources other than a laser, so there is no interference.

Volumetric Techniques

Medical

Computed tomography (CT) is a medical imaging method which generates a three-dimensional image of the inside of an object from a large series of two-dimensional X-ray images, similarly Magnetic resonance imaging is another medical imaging technique that provides much greater contrast between the different soft tissues of the body than computed tomography (CT) does, making it especially useful in neurological (brain), musculoskeletal, cardiovascular, and oncological (cancer) imaging. These techniques produce a discrete 3D volumetric representation that can be directly visualised, manipulated or converted to traditional 3D surface by mean of isosurface extraction algorithms.

Industrial

Although most common in medicine, Industrial computed tomography, Microtomography and MRI are also used in other fields for acquiring a digital representation of an object and its interior, such as non destructive materials testing, reverse engineering, or studying biological and paleontological specimens.

Non-Contact Passive

Passive 3D imaging solutions do not emit any kind of radiation themselves, but instead rely on detecting reflected ambient radiation. Most solutions of this type detect visible light because it is a readily available ambient radiation. Other types of radiation, such as infra red could also be used. Passive methods can be very cheap, because in most cases they do not need particular hardware but simple digital cameras.

- *Stereoscopic systems usually employ two video cameras, slightly apart, looking at the same scene. By analysing the slight differences between the images* *seen by each camera, it is possible to determine the distance at each point in the images. This method is based on the same principles driving human ste-*reoscopic vision.

- *Photometric systems usually use a single camera, but take multiple images under varying lighting conditions. These techniques attempt to invert the image formation model in order to recover the surface orientation at each pixel.*

- *Silhouette techniques use outlines created from a sequence of photographs around a three-dimensional object against a well contrasted background. These* silhouettes are extruded and intersected to form the visual hull approximation of the object. With these approaches some concavities of an object (like the interior of a bowl) cannot be detected.

User Assisted (Image-Based Modelling)

There are other methods that, based on the user assisted detection and identification of some features and shapes on a set of different pictures of an object are able to build an approximation of the object itself. This kind of techniques are useful to build fast approximation of simple shaped objects like buildings. Various commercial packages are available like D-Sculptor, iModeller, Autodesk ImageModeler, 123DCatch or PhotoModeler.

This sort of 3D imaging solution is based on the principles of photogrammetry. It is also somewhat similar in methodology to panoramic photography, except that the photos are taken of one object on a three-dimensional space in order to replicate it instead of taking a series of photos from one point in a three-dimensional space in order to replicate the surrounding environment.

Reconstruction

From Point Clouds

The point clouds produced by 3D scanners and 3D imaging can be used directly for measurement and visualisation in the architecture and construction world.

From Models

Most applications, however, use instead polygonal 3D models, NURBS surface models, or editable feature-based CAD models (aka Solid models).

- Polygon mesh models: In a polygonal representation of a shape, a curved surface is modeled as many small faceted flat surfaces (think of a sphere modeled as a disco ball). Polygon models—also called Mesh models, are useful for visualisation, for some CAM (i.e., machining), but are generally "heavy" (i.e., very large data sets), and are relatively un-editable in this form. Reconstruction to polygonal model involves finding and connecting adjacent points with straight lines in order to create a continuous surface. Many applications, both free and nonfree, are available for this purpose (e.g. MeshLab, PointCab, kubit Point-Cloud for AutoCAD, JRC 3D Reconstructor, imagemodel, PolyWorks, Rapidform, Geomagic, Imageware, Rhino 3D etc.).

- Surface models: The next level of sophistication in modeling involves using a quilt of *curved surface patches to model our shape. These might be NURBS, TSplines or other curved representations of curved topology. Using NURBS, our sphere is a true mathematical sphere. Some applications offer patch layout by hand but the best in class offer both automated patch layout and manual layout. These patches have the advantage of being lighter and more manipulable when exported to CAD. Surface models are somewhat editable, but only in a sculptural sense of pushing and pulling to deform the surface. This representation lends itself well to modelling organic and artistic shapes. Providers of surface modellers include Rapidform,* Geomagic, Rhino 3D, Maya, T Splines etc.

- Solid CAD models: From an engineering/manufacturing perspective, the ultimate representation of a digitised shape is the editable, parametric CAD model. After all, CAD is the common "language" of industry to describe, edit and maintain the shape of the enterprise's assets. In CAD, our sphere is described by parametric features which are easily edited by changing a value (e.g., centre point and radius).

These CAD models describe not simply the envelope or shape of the object, but CAD models also embody the "design intent" (i.e., critical features and their relationship to other features). An example of design intent not evident in the shape alone might be a brake drum's lug bolts, which must be concentric with the hole in the centre of the drum. This knowledge would drive the sequence and method of creating the CAD model; a designer with an awareness of this relationship would not design the lug bolts referenced to the outside diameter, but instead, to the center. A modeler creating a CAD model will want to include both Shape and design intent in the complete CAD model.

Vendors offer different approaches to getting to the parametric CAD model. Some export the NURBS surfaces and leave it to the CAD designer to complete the model in CAD (e.g., Geomagic, Imageware, Rhino 3D). Others use the scan data to create an editable and verifiable feature based model that is imported into CAD with full feature tree intact, yielding a complete, native CAD model, capturing both shape and design intent (e.g. Geomagic, Rapidform). Still other CAD applications are robust enough to manipulate limited points or polygon models within the CAD environment (e.g., CATIA, AutoCAD, Revit).

From a Set of 2D Slices

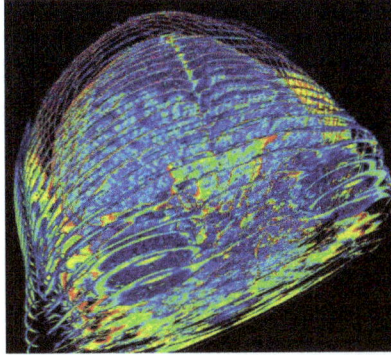

3D reconstruction of the brain and eyeballs from CT scanned DICOM images. In this image, areas with the density of bone or air were made transparent, and the slices stacked up in an approximate free-space alignment. The outer ring of material around the brain are the soft tissues of skin and muscle on the outside of the skull. A black box encloses the slices to provide the black background. Since these are simply 2D images stacked up, when viewed on edge the slices disappear since they have effectively zero thickness. Each DICOM scan represents about 5mm of material averaged into a thin slice.

CT, industrial CT, MRI, or Micro-CT scanners do not produce point clouds but a set of 2D slices (each termed a "tomogram") which are then 'stacked together' to produce a 3D representation. There are several ways to do this depending on the output required:

- Volume rendering: Different parts of an object usually have different threshold values or greyscale densities. From this, a 3-dimensional model can be constructed and displayed on screen. Multiple models can be constructed from various thresholds, allowing different colours to represent each component of the object. Volume rendering is usually only used for visualisation of the scanned object.

- Image segmentation: Where different structures have similar threshold/greyscale values, it can become impossible to separate them simply by adjusting volume rendering parameters. The solution is called segmentation, a manual or automatic procedure that can remove the unwanted structures from the image. Image segmentation software usually allows export of the segmented structures in CAD or STL format for further manipulation.

- Image-based meshing: When using 3D image data for computational analysis (e.g. CFD and FEA), simply segmenting the data and meshing from CAD can become time consuming, and virtually intractable for the complex topologies typical of image data. The solution is called image-based meshing, an automated process of generating an accurate and realistic geometrical description of the scan data.

From Laser Scans

Laser scanning describes the general method to sample or scan a surface using laser technology. Several areas of application exist that mainly differ in the power of the lasers that are used, and in the results of the scanning process. Low laser power is used when the scanned surface doesn't have to be influenced, e.g. when it only has to be digitised. Confocal or 3D laser scanning are methods to get information about the scanned surface. Another low-power application uses structured light projection systems for solar cell flatness metrology, enabling stress calculation throughout in excess of 2000 wafers per hour.

The laser power used for laser scanning equipment in industrial applications is typically less than 1W. The power level is usually on the order of 200 mW or less but sometime more.

Applications

Construction industry and Civil Engineering

- Robotic control: e.g. a laser scanner may function as the "eye" of a robot.
- As-built drawings of bridges, industrial plants, and monuments
- Documentation of historical sites
- Site modelling and lay outing
- Quality control
- Quantity surveys
- Freeway redesign
- Establishing a bench mark of pre-existing shape/state in order to detect structural changes resulting from exposure to extreme loadings such as earthquake, vessel/truck impact or fire.
- Create GIS (geographic information system) maps and geomatics.
- Subsurface laser scanning in mines and Karst voids.
- Forensic documentation

Design Process

- Increasing accuracy working with complex parts and shapes,
- Coordinating product design using parts from multiple sources,
- Updating old CD scans with those from more current technology,
- Replacing missing or older parts,
- Creating cost savings by allowing as-built design services, for example in automotive manufacturing plants,

- "Bringing the plant to the engineers" with web shared scans, and
- Saving travel costs.

Entertainment

3D scanners are used by the entertainment industry to create digital 3D models for movies, video games and leisure purposes. They are heavily utilised in virtual cinematography. In cases where a real-world equivalent of a model exists, it is much faster to scan the real-world object than to manually create a model using 3D modeling software. Frequently, artists sculpt physical models of what they want and scan them into digital form rather than directly creating digital models on a computer.

Law Enforcment

3D laser scanning is used by the FBI. 3D Models are used for on-site documentation of:

- Crime scenes
- Bullet trajectory
- Accident reconstruction
- Bombings
- Plane crashes, and more

Reverse Engineering

Reverse engineering of a mechanical component requires a precise digital model of the objects to be reproduced. Rather than a set of points a precise digital model can be represented by a polygon mesh, a set of flat or curved NURBS surfaces, or ideally for mechanical components, a CAD solid model. A 3D scanner can be used to digitise free-form or gradually changing shaped components as well as prismatic geometries whereas a coordinate measuring machine is usually used only to determine simple dimensions of a highly prismatic model. These data points are then processed to create a usable digital model, usually using specialized reverse engineering software.

Cultural Heritage

There have been many research projects undertaken via the scanning of historical sites and artifacts both for documentation and analysis purposes.

The combined use of 3D scanning and 3D printing technologies allows the replication of real objects without the use of traditional plaster casting techniques, that in many cases can be too invasive for being performed on precious or delicate cultural heritage artifacts. In an example of a typical application scenario, a gargoyle model was digitally acquired using a 3D scanner and the produced 3D data was processed using MeshLab. The resulting digital 3D model was fed to a rapid prototyping machine to create a real resin replica of the original object.

Michelangelo

In 1999, two different research groups started scanning Michelangelo's statues. Stanford University with a group led by Marc Levoy used a custom laser triangulation scanner built by Cyberware to scan Michelangelo's statues in Florence, notably the David, the Prigioni and the four statues in The Medici Chapel. The scans produced a data point density of one sample per 0.25 mm, detailed enough to see Michelangelo's chisel marks. These detailed scans produced a large amount of data (up to 32 gigabytes) and processing the data from his scans took 5 months. Approximately in the same period a research group from IBM, led by H. Rushmeier and F. Bernardini scanned the Pietà of Florence acquiring both geometric and colour details. The digital model, result of the Stanford scanning campaign, was thoroughly used in the 2004 subsequent restoration of the statue.

Monticello

In 2002, David Luebke, et al. scanned Thomas Jefferson's Monticello. A commercial time of flight laser scanner, the DeltaSphere 3000, was used. The scanner data was later combined with colour data from digital photographs to create the Virtual Monticello, and the Jefferson's Cabinet exhibits in the New Orleans Museum of Art in 2003. The Virtual Monticello exhibit simulated a window looking into Jefferson's Library. The exhibit consisted of a rear projection display on a wall and a pair of stereo glasses for the viewer. The glasses, combined with polarised projectors, provided a 3D effect. Position tracking hardware on the glasses allowed the display to adapt as the viewer moves around, creating the illusion that the display is actually a hole in the wall looking into Jefferson's Library. The Jefferson's Cabinet exhibit was a barrier stereogram (essentially a non-active hologram that appears different from different angles) of Jefferson's Cabinet.

Cuneiform Tablets

In 2003, Subodh Kumar, et al. undertook the 3D scanning of ancient cuneiform tablets. Again, a laser triangulation scanner was used. The tablets were scanned on a regular grid pattern at a resolution of 0.025 mm (0.00098 in).

Kasubi Tombs

A 2009 CyArk 3D scanning project at Uganda's historic Kasubi Tombs, a UNESCO World Heritage Site, using a Leica HDS 4500, produced detailed architectural models of Muzibu Azaala Mpanga, the main building at the complex and tomb of the Kabakas (Kings) of Uganda. A fire on March 16, 2010, burned down much of the Muzibu Azaala Mpanga structure, and reconstruction work is likely to lean heavily upon the dataset produced by the 3D scan mission.

"Plastico Di Roma Antica"

In 2005, Gabriele Guidi, et al. scanned the "Plastico di Roma antica", a model of Rome created in the last century. Neither the triangulation method, nor the time of flight method satisfied the requirements of this project because the item to be scanned was both large and contained small details. They found though, that a modulated light scanner was able to provide both the ability to scan an object the size of the model and the accuracy that was needed. The modulated light scanner was supplemented by a triangulation scanner which was used to scan some parts of the model.

Other Projects

The 3D Encounters Project at the Petrie Museum of Egyptian Archaeology aims to use 3D laser scanning to create a high quality 3D image library of artefacts and enable digital travelling exhibitions of fragile Egyptian artefacts, English Heritage has investigated the use of 3D laser scanning for a wide range of applications to gain archaeological and condition data, and the National Conservation Centre in Liverpool has also produced 3D laser scans on commission, including portable object and in situ scans of archaeological sites. The Smithsonian Institution has a project called Smithsonian X 3D notable for the breadth of types of 3D objects they are attempting to scan. These include small objects such as insects and flowers, to human sized objects such as Amelia Earhart's Flight Suit to room sized objects such as the Gunboat Philadelphia to historic sites such as Liang Bua in Indonesia. Also of note the data from these scans is being made available to the public for free and downloadable in several data formats.

Medical CAD/CAM

3D scanners are used to capture the 3D shape of a patient in orthotics and dentistry. It gradually supplants tedious plaster cast. CAD/CAM software are then used to design and manufacture the orthosis, prosthesis or dental implants.

Many Chairside dental CAD/CAM systems and Dental Laboratory CAD/CAM systems use 3D Scanner technologies to capture the 3D surface of a dental preparation (either *in vivo or in vitro), in order to produce a restoration digitally using CAD software and ultimately produce the final restoration using a CAM technology (such as a CNC milling machine, or 3D printer). The chairside systems are designed to facilitate the 3D scanning of a preparation in vivo and produce the restoration (such as a Crown, Onlay, Inlay or Veneer).*

Quality Assurance and Industrial Metrology

The digitalisation of real-world objects is of vital importance in various application domains. This method is especially applied in industrial quality assurance to measure the geometric dimension accuracy. Industrial processes such as assembly are complex, highly automated and typically based on CAD (Computer Aided Design) data. The problem is that the same degree of automation is also required for quality assurance. It is, for example, a very complex task to assemble a modern car, since it consists of many parts that must fit together at the very end of the production line. The optimal performance of this process is guaranteed by quality assurance systems. Especially the geometry of the metal parts must be checked in order to assure that they have the correct dimensions, fit together and finally work reliably.

Within highly automated processes, the resulting geometric measures are transferred to machines that manufacture the desired objects. Due to mechanical uncertainties and abrasions, the result may differ from its digital nominal. In order to automatically capture and evaluate these deviations, the manufactured part must be digitised as well. For this purpose, 3D scanners are applied to generate point samples from the object's surface which are finally compared against the nominal data.

The process of comparing 3D data against a CAD model is referred to as CAD-Compare, and can be a useful technique for applications such as determining wear patterns on moulds and tooling, determining accuracy of final build, analysing gap and flush, or analysing highly complex sculpted surfaces. At present, laser triangulation scanners, structured light and contact scanning are the predominant technologies employed for industrial purposes, with contact scanning remaining the slowest, but overall most accurate option.

Barcode Reader

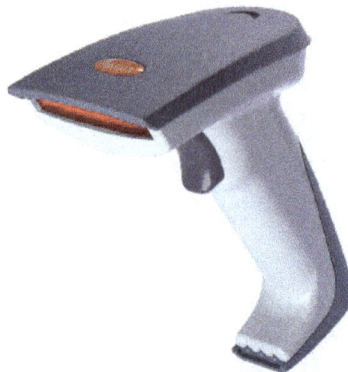

A handheld barcode scanner

A barcode reader (or barcode scanner) is an electronic device that can read and output printed barcodes to a computer. Like a flatbed scanner, it consists of a light source, a lens and a light sensor translating optical impulses into electrical ones. Additionally, nearly all barcode readers contain *decoder circuitry analyzing the barcode's image data provided by the sensor and sending the barcode's content to the scanner's output port.*

A stationary barcode scanner for a conveyor line

Types of Barcode Scanners

Technology

Barcode readers can be differentiated by technologies as follows:

Pen-type scanners Pen-type readers consist of a light source and photodiode that are placed next to each other in the tip of a pen or wand. To read a bar code, the person holding the pen must move the tip of it across the bars at a relatively uniform speed. The photodiode measures the intensity of the light reflected back from the light source as the tip crosses each bar and space in the printed code. The photodiode generates a waveform that is used to measure the widths of the bars and spaces in the bar code. Dark bars in the bar code absorb light and white spaces reflect light so that the voltage waveform generated by the photodiode is a representation of the bar and space pattern in the bar code. This waveform is decoded by the scanner in a manner similar to the way Morse code dots and dashes are decoded.

Laser Scanners

Laser scanners work the same way as pen type readers except that they use a laser beam as the light source and typically employ either a reciprocating mirror or a rotating prism to scan the laser beam back and forth across the bar code. As with the pen type reader, a photo-diode is used to measure the intensity of the light reflected back from the bar code. In both pen readers and laser scanners, the light emitted by the reader is rapidly varied in brightness with a data pattern and the photo-diode receive circuitry is designed to detect only signals with the same modulated pattern.

CCD readers (Also Known as LED Scanners)

CCD readers use an array of hundreds of tiny light sensors lined up in a row in the head of the reader. Each sensor measures the intensity of the light immediately in front of it. Each individual light sensor in the CCD reader is extremely small and because there are hundreds of sensors lined up in a row, a voltage pattern identical to the pattern in a bar code is generated in the reader by sequentially measuring the voltages across each sensor in the row. The important difference between a CCD reader and a pen or laser scanner is that the CCD reader is measuring emitted ambient light from the bar code whereas pen or laser scanners are measuring reflected light of a specific frequency originating from the scanner itself.

Camera-Based Readers

Two-dimensional imaging scanners are a newer type of bar code reader. They use a camera and image processing techniques to decode the bar code.

vVideo camera readers use small video cameras with the same CCD technology as in a CCD bar code reader except that instead of having a single row of sensors, a video camera has hundreds of rows of sensors arranged in a two dimensional array so that they can generate an image.

Large field-of-view readers use high resolution industrial cameras to capture multiple bar codes simultaneously. All the bar codes appearing in the photo are decoded instantly (ImageID patents and code creation tools) or by use of plugins (e.g. the Barcodepedia used a flash application and some web cam for querying a database), have been realized options for resolving the given tasks.

Omnidirectional Barcode Scanners

Omnidirectional scanning uses "series of straight or curved scanning lines of varying directions in the form of a starburst, a Lissajous pattern, or other multiangle arrangement are projected at the symbol and one or more of them will be able to cross all of the symbol's bars and spaces, no matter what the orientation."

Omnidirectional scanners almost all use a laser. Unlike the simpler single-line laser scanners, they produce a pattern of beams in varying orientations allowing them to read barcodes presented to it at different angles. Most of them use a single rotating polygonal mirror and an arrangement of several fixed mirrors to generate their complex scan patterns.

Omnidirectional scanners are most familiar through the horizontal scanners in supermarkets, where packages are slid over a glass or sapphire window. There are a range of different omnidirectional units available which can be used for differing scanning applications, ranging from retail type applications with the barcodes read only a few centimetres away from the scanner to industrial conveyor scanning where the unit can be a couple of metres away or more from the code. Omnidirectional scanners are also better at reading poorly printed, wrinkled, or even torn barcodes.

Cell Phone Cameras

While cell phone cameras without auto-focus are not ideal for reading some common barcode formats, there are 2D barcodes which are optimized for cell phones, as well as QR Codes and Data Matrix codes which can be read quickly and accurately with or without auto-focus.

Cell phone cameras open up a number of applications for consumers:

- Movies: DVD/VHS movie catalogs.

- Music: CD catalogs – play MP3 when scanned.

- Book catalogs and device.

- Groceries, nutrition information, making shopping lists when the last of an item is used, etc.

- Personal Property inventory (for insurance and other purposes)ode scanned into personal finance software when entering. Later, scanned receipt images can then be automatically associated with the appropriate entries. Later, the bar codes can be used to rapidly weed out paper copies not required to be retained for tax or asset inventory purposes.

- If retailers put barcodes on receipts that allowed downloading an electronic copy or encoded the entire receipt in a 2D barcode, consumers could easily import data into personal finance, property inventory, and grocery management software. Receipts scanned on a scanner could be automatically identified and associated with the appropriate entries in finance and property inventory software.

- Consumer tracking from the retailer perspective (for example, loyalty card programs that track consumers purchases at the point of sale by having them scan a QR code).

A number of enterprise applications using cell phones are appearing:

- Access control (for example, ticket validation at venues), inventory reporting (for example, tracking deliveries), asset tracking (for example, anti-counterfeiting).

Smartphones

- Smartphones can be used in Google's mobile Android operating system via both their own Google Goggles application. Nokia's Symbian operating system features a barcode scanner which can scan barcodes, while mbarcode is a barcode reader for the Maemo operating system. In the Apple iOS, a barcode reader is not automatically included, but there are more than fifty free or paid apps available with both scanning capabilities and hard-linking to URI. With BlackBerry devices, the App World application can natively scan barcodes. Windows Phone 8 is able to scan barcodes through the Bing search app.

Housing

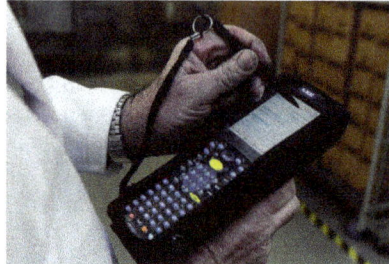

A large multifunction barcode scanner being used to monitor the transportation of packages of radioactive pharmaceuticals

Barcode readers can be distinguished based on housing design as follows:

Handheld scanner

> with a handle and typically a trigger button for switching on the light source.

Pen scanner (or wand scanner)

> a pen-shaped scanner that is swiped.

Stationary scanner

> Wall- Or Table-Mounted Scanners That The Barcode Is Passed Under Or Beside. These Are Commonly Found at the checkout counters of supermarkets and other retailers.

Fixed-position scanner

> an industrial barcode reader, used to identify products during manufacture or logistics. Often used on conveyor tracks to identify cartons or pallets which need to be routed to another process or shipping location. Another application joins holographic scanners with a checkweigher to read bar codes of any orientation or placement, and weighs the package. Systems like this are used in factory and farm automation for quality management and shipping.

PDA scanner (or Auto-ID PDA)

> a PDA with a built-in barcode reader or attached barcode scanner.

Automatic reader

> a back office equipment to read barcoded documents at high speed (50,000/ hour).

Cordless scanner (or Wireless scanner)

> a cordless barcode scanner is operated by a battery fit inside it and is not connected to the electricity mains and transfer data to the connected device like PC.

Methods of Connection

Early Serial Interfaces

Early barcode scanners, of all formats, almost universally used the then-common RS-232 serial interface. This was an electrically simple means of connection and the software to access it is also relatively simple, although needing to be written for specific computers and their serial ports.

Proprietary Interfaces

There are a few other less common interfaces. These were used in large EPOS systems with dedicated hardware, rather than attaching to existing commodity computers. In some of these interfaces, the scanning device returned a "raw" signal proportional to the intensities seen while scanning the barcode. This was then decoded by the host device. In some cases the scanning device would convert the symbology of the barcode to one that could be recognized by the host device, such as Code 39.

Keyboard Wedge (e.g. PS/2)

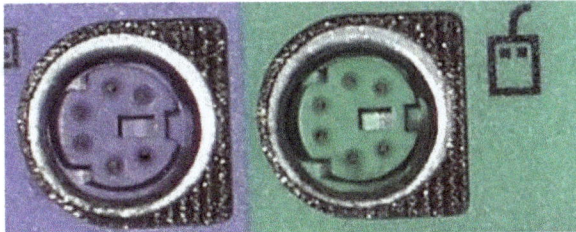

PS/2 keyboard and mouse ports

With the popularity of the PC and its standard keyboard interface, it became ever easier to connect physical hardware to a PC and so there was commercial demand similarly to reduce the complexity of the associated software. "Keyboard wedge" (PS/2) hardware plugged between the PC and its normal keyboard, with characters from the barcode scanner appearing exactly as if they had been typed at the keyboard. This made the addition of simple barcode reading abilities to existing programs very easy, without any need to change them, although it did require some care by the user and could be restrictive in the content of the barcodes that could be handled.

USB

Later barcode readers began to use USB connectors rather than the keyboard port, as this became a more convenient hardware option. To retain the easy integration with existing programs, a device driver called a "software wedge" could be used, to emulate the keyboard-impersonating behavior of the old "keyboard wedge" hardware.

In many cases, a choice of USB interface types (HID, CDC) are provided. Some have PoweredUSB.

Wireless Networking

Some modern handheld barcode readers can be operated in wireless networks according to IEEE 802.11g (WLAN) or IEEE 802.15.1 (Bluetooth). Some barcode readers also support radio frequencies viz. 433 MHz or 910 MHz. Readers without external power sources require their batteries be recharged occasionally, which may make them unsuitable for some uses.

Resolution

The scanner resolution is measured by the size of the dot of light emitted by the reader. If this dot of light is wider than any bar or space in the bar code, then it will overlap two elements (two spaces or two bars) and it may produce wrong output. On the other hand, if a too small dot of light is used, then it can misinterpret any spot on the bar code making the final output wrong.

The most commonly used dimension is 13 thou (0.013 in or 0.33 mm), although some scanners can read codes with dimensions as small as 3 thou (0.003 in or 0.075 mm). Most manufacturers advertise bar code resolution in *mil, which is interchangeable with thou. Smaller bar codes must be printed at high resolution to be read accurately.*

Laser surgery

Laser surgery is surgery using a laser (instead of a scalpel) to cut tissue. Examples include the use of a laser scalpel in otherwise conventional surgery, and soft-tissue laser surgery, in which the laser beam vaporizes soft tissue with high water content. Laser resurfacing is a technique in which covalent bonds of a material are dissolved by a laser, a technique invented by aesthetic plastic surgeon Thomas L Roberts, III using CO_2 lasers in the 1990s. The CO_2 (carbon dioxide) laser remains the gold standard for the soft tissue surgery because of the ease of simultaneous photo-thermal ablation and coagulation (and small blood capillary hemostasis).

Laser surgery is commonly used on the eye. Techniques used include LASIK, which is used to correct near and far-sightedness in vision, and photorefractive keratectomy, a procedure which permanently reshapes the cornea using an excimer laser to remove a small amount of tissue. Types of surgical lasers include carbon dioxide, argon, Nd:YAG laser, and Potassium titanyl phosphate.

Effects

1. Photochemical effect: clinically referred to as photodynamic therapy. Photo-sensitizer (photophrin II) is administered which is taken up by the tumor tissue and later irradiated by laser light resulting in highly toxic substances with resultant necrosis of the tumor. Photodynamic therapy is used in palliation of oesophagial and bronchial carcinoma and ablation of mucosal cancers of Gastrointestinal tract and urinary bladder.

2. Photoablative effect: Used in eye surgeries like band keratoplast, and endartectomy of peripheral blood vessels.

3. Photothermal effect: this property is used for endoscopic control of bleeding e.g. Bleeding peptic ulcers, oesophagial varices

4. Photomechanical effect: used in intraluminal lithotripsy

Applications

Dermatology and Plastic Surgery

A range of lasers such as erbium, dye, and CO2 are used to treat various skin conditions including scars, vascular and pigmented lesions, and for photorejuvenation.

Eye Surgery

Various types of laser surgery are used to treat refractive error:

- ReLEx SMILE, use a femtosecond laser to create a refractive lenticule within the stroma which is then removed through a small incision

- LASIK, in which a knife is used to cut a flap in the cornea, and a laser is used to reshape the layers underneath, to treat refractive error

- IntraLASIK, a variant in which the flap is also cut with a laser

- Photorefractive keratectomy (PRK, LASEK), in which the cornea is reshaped without first cutting a flap

- Laser thermal keratoplasty, in which a ring of concentric burns is made in the cornea, which cause its surface to steepen, allowing better near vision

Lasers are also used to treat non-refractive conditions, such as:

- Phototherapeutic keratectomy (PTK), in which opacities and surface irregularities are removed from the cornea

- Laser coagulation, in which a laser is used to cauterize blood vessels in the eye, to treat various conditions

- Lasers can be used to repair tears in the retina.

Endovascular Surgery

Laser endarterectomy is a technique in which an entire atheromatous plaque in the artery is excised. Laser recanalization of blocked arteries. other applications include laser assisted angioplasties and laser assisted vascular anastamosis.

Foot and Ankle Surgery

Lasers are used to treat several disorders in foot and ankle surgery. They are used to remove benign and malignant tumors, treat bunions, debride ulcers and burns, excise epidermal nevi, blue rubber bleb nevi, and keloids, and the removal of hypertrophic scars and tattoos.

A carbon dioxide laser (CO2) is used in surgery to treat onychocryptosis (ingrown nails), onychauxis (club nails), onychogryposis (rams horn nail), and onychomycosis (fungus nail).

Gastro-Intestinal Tract

- Peritoneum-Laser is used for adhesiolysis.

- Peptic ulcer disease and oesophageal varices - Laser photoablation is done.

- Coagulation of vascular malformations of stomach, duodenum and colon.

- Lasers can be effectively used to treat early gastric cancers provided they are less than 4 cm and without lymph node involvement. Lasers are also used in treating oral submucous fibrosis.

- Palliative laser therapy is given in advanced oesophageal cancers with obstruction of lumen. Recanalisation of the lumen is done which allows the patient to resume soft diet and maintain hydration.

- Ablative laser therapy is used in advanced colorectal cancers to relieve obstruction and to control bleeding.

- Laser surgery used in hemorrhoidectomy, and is a relatively popular and non-invasive method of hemorrhoid removal.

- Laser-assisted liver resections have been done using carbon dioxide and Nd:YAG lasers.

- Ablation of liver tumors can be achieved by selective photovaporization of the tumor.

- Endoscopic laser lithotripsy is a safer modality compared to electrohydraulic lithotripsy.

Oral and Dental Surgery

The CO_2 laser is used in oral and dental surgery for virtually all soft-tissue procedures, such as gingivecomies, vestibuloplasties, frenectomies and operculectomies. The CO_2 10,600 nm wavelength is safe around implants as it is reflected by titanium, and thus has been gaining popularity in the field of periodontology. The laser may also be effective in treating peri-implantitis.

Other Surgery

The CO_2 laser is also used in gynecology, genitourinary, general and thoracic surgery, otorhinolaryngology, orthopedic, and neurosurgery.

Fiber-Optic Communication

An optical fiber junction box. The yellow cables are single mode fibers; the orange and blue cables are multi-mode fibers: 62.5/125 μm OM1 and 50/125 μm OM3 fibers, respectively.

Stealth installing a 432-count dark fibre cable underneath the streets of Midtown Manhattan, New York City

Fiber-optic communication is a method of transmitting information from one place to another by sending pulses of light through an optical fiber. The light forms an electromagnetic carrier wave that is modulated to carry information. First developed in the 1970s, fiber-optics have revolutionized the telecommunications industry and have played a major role in the advent of the Information Age. Because of its advantages over electrical transmission, optical fibers have largely replaced copper wire communications in core networks in the developed world. Optical fiber is used by many telecommunications companies to transmit telephone signals, Internet communication, and cable television signals. Researchers at Bell Labs have reached internet speeds of over 100 petabit×kilometer per second using fiber-optic communication.

The process of communicating using fiber-optics involves the following basic steps:

- creating the optical signal involving the use of a transmitter, usually from an electrical signal

- relaying the signal along the fiber, ensuring that the signal does not become too distorted or weak

- receiving the optical signal

- converting it into an electrical signal

Applications

Optical fiber is used by many telecommunications companies to transmit telephone signals, Internet communication, and cable television signals. Due to much lower attenuation and interference, optical fiber has large advantages over existing copper wire in long-distance and high-demand applications. However, infrastructure development within cities was relatively difficult and time-consuming, and fiber-optic systems were complex and expensive to install and operate. Due to these difficulties, fiber-optic communication systems have primarily been installed in long-distance applications, where they can be used to their full transmission capacity, offsetting the increased cost. Since 2000, the prices for fiber-optic communications have dropped considerably.

The price for rolling out fiber to the home has currently become more cost-effective than that of rolling out a copper based network. Prices have dropped to $850 per subscriber in the US and lower in countries like The Netherlands, where digging costs are low and housing density is high.

Since 1990, when optical-amplification systems became commercially available, the telecommunications industry has laid a vast network of intercity and transoceanic fiber communication lines. By 2002, an intercontinental network of 250,000 km of submarine communications cable with a capacity of 2.56 Tb/s was completed, and although specific network capacities are privileged information, telecommunications investment reports indicate that network capacity has increased dramatically since 2004.

History

In 1880 Alexander Graham Bell and his assistant Charles Sumner Tainter created a very early precursor to fiber-optic communications, the Photophone, at Bell's newly established Volta Laboratory in Washington, D.C. Bell considered it his most important invention. The device allowed for the transmission of sound on a beam of light. On June 3, 1880, Bell conducted the world's first wireless telephone transmission between two buildings, some 213 meters apart. Due to its use of an atmospheric transmission medium, the Photophone would not prove practical until advances in laser and optical fiber technologies permitted the secure transport of light. The Photophone's first practical use came in military communication systems many decades later.

In 1954 Harold Hopkins and Narinder Singh Kapany showed that rolled fiber glass allowed light to be transmitted. Initially it was considered that the light can traverse in only straight medium.

In 1966 Charles K. Kao and George Hockham proposed optical fibers at STC Laboratories (STL) at Harlow, England, when they showed that the losses of 1,000 dB/km in existing glass (compared to 5-10 dB/km in coaxial cable) was due to contaminants, which could potentially be removed.

Optical fiber was successfully developed in 1970 by Corning Glass Works, with attenuation low enough for communication purposes (about 20dB/km), and at the same time GaAs semiconductor lasers were developed that were compact and therefore suitable for transmitting light through fiber optic cables for long distances.

After a period of research starting from 1975, the first commercial fiber-optic communications system was developed, which operated at a wavelength around 0.8 μm and used GaAs semiconductor lasers. This first-generation system operated at a bit rate of 45 Mbps with repeater spacing of up to 10 km. Soon on 22 April 1977, General Telephone and Electronics sent the first live telephone traffic through fiber optics at a 6 Mbit/s throughput in Long Beach, California.

In October 1973, Corning Glass signed a development contract with CSELT and Pirelli aimed to test fiber optics in an urban environment: in September 1977, the second cable in this test series, named COS-2, was experimentally deployed in two lines (9 km) in Turin, for the first time in a big city, at a speed of 140 Mbit/s.

The second generation of fiber-optic communication was developed for commercial use in the early 1980s, operated at 1.3 μm, and used InGaAsP semiconductor lasers. These early systems were initially limited by multi mode fiber dispersion, and in 1981 the single-mode fiber was revealed to greatly improve system performance, however practical connectors capable of working with single mode fiber proved difficult to develop. In 1984, they had already developed a fiber optic cable that would help further their progress toward making fiber optic cables that would circle the globe. Canadian service provider SaskTel had completed construction of what was then the world's longest commercial fiberoptic network, which covered 3,268 km and linked 52 communities. By 1987, these systems were operating at bit rates of up to 1.7 Gb/s with repeater spacing up to 50 km.

The first transatlantic telephone cable to use optical fiber was TAT-8, based on Desurvire optimized laser amplification technology. It went into operation in 1988.

Third-generation fiber-optic systems operated at 1.55 µm and had losses of about 0.2 dB/km. This development was spurred by the discovery of Indium gallium arsenide and the development of the Indium Gallium Arsenide photodiode by Pearsall. Engineers overcame earlier difficulties with pulse-spreading at that wavelength using conventional InGaAsP semiconductor lasers. Scientists overcame this difficulty by using dispersion-shifted fibers designed to have minimal dispersion at 1.55 µm or by limiting the laser spectrum to a single longitudinal mode. These developments eventually allowed third-generation systems to operate commercially at 2.5 Gbit/s with repeater spacing in excess of 100 km.

The fourth generation of fiber-optic communication systems used optical amplification to reduce the need for repeaters and wavelength-division multiplexing to increase data capacity. These two improvements caused a revolution that resulted in the doubling of system capacity every six months starting in 1992 until a bit rate of 10 Tb/s was reached by 2001. In 2006 a bit-rate of 14 Tbit/s was reached over a single 160 km line using optical amplifiers.

The focus of development for the fifth generation of fiber-optic communications is on extending the wavelength range over which a WDM system can operate. The conventional wavelength window, known as the C band, covers the wavelength range 1.53-1.57 µm, and *dry fiber has a low-loss window promising an extension of that range to 1.30-1.65 µm. Other developments include the concept of* "optical solitons, " pulses that preserve their shape by counteracting the effects of dispersion with the nonlinear effects of the fiber by using pulses of a specific shape.

In the late 1990s through 2000, industry promoters, and research companies such as KMI, and RHK predicted massive increases in demand for communications bandwidth due to increased use of the Internet, and commercialization of various bandwidth-intensive consumer services, such as video on demand. Internet protocol data traffic was increasing exponentially, at a faster rate than integrated circuit complexity had increased under Moore's Law. From the bust of the dot-com bubble through 2006, however, the main trend in the industry has been consolidation of firms and offshoring of manufacturing to reduce costs. Companies such as Verizon and AT&T have taken advantage of fiber-optic communications to deliver a variety of high-throughput data and broadband services to consumers' homes.

Technology

Modern fiber-optic communication systems generally include an optical transmitter to convert an electrical signal into an optical signal to send into the optical fiber, a cable containing bundles of multiple optical fibers that is routed through underground conduits and buildings, multiple kinds of amplifiers, and an optical receiver to recover the signal as an electrical signal. The information transmitted is typically digital information generated by computers, telephone systems, and cable television companies.

Transmitters

A GBIC module (shown here with its cover removed), is an optical and electrical transceiver. The electrical connector is at top right, and the optical connectors are at bottom left

The most commonly used optical transmitters are semiconductor devices such as light-emitting diodes (LEDs) and laser diodes. The difference between LEDs and laser diodes is that LEDs produce incoherent light, while laser diodes produce coherent light. For use in optical communications, semiconductor optical transmitters must be designed to be compact, efficient, and reliable, while operating in an optimal wavelength range, and directly modulated at high frequencies.

In its simplest form, a LED is a forward-biased p-n junction, emitting light through spontaneous emission, a phenomenon referred to as electroluminescence. The emitted light is incoherent with a relatively wide spectral width of 30-60 nm. LED light transmission is also inefficient, with only about 1% of input power, or about 100 microwatts, eventually converted into launched power which has been coupled into the optical fiber. However, due to their relatively simple design, LEDs are very useful for low-cost applications.

Communications LEDs are most commonly made from Indium gallium arsenide phosphide (InGaAsP) or gallium arsenide (GaAs). Because InGaAsP LEDs operate at a longer wavelength than GaAs LEDs (1.3 micrometers vs. 0.81-0.87 micrometers), their output spectrum, while equivalent in energy is wider in wavelength terms by a factor of about 1.7. The large spectrum width of LEDs is subject to higher fiber dispersion, considerably limiting their bit rate-distance product (a common measure of usefulness). LEDs are suitable primarily for local-area-network applications with bit rates of 10-100 Mbit/s and transmission distances of a few kilometers. LEDs have also been developed that use several quantum wells to emit light at different wavelengths over a broad spectrum, and are currently in use for local-area WDM (Wavelength-Division Multiplexing) networks.

Today, LEDs have been largely superseded by VCSEL (Vertical Cavity Surface Emitting Laser) devices, which offer improved speed, power and spectral properties, at a similar cost. Common VCSEL devices couple well to multi mode fiber.

A semiconductor laser emits light through stimulated emission rather than spontaneous emission, which results in high output power (~100 mW) as well as other benefits related to the nature of coherent light. The output of a laser is relatively directional, allowing high coupling efficiency (~50 %) into single-mode fiber. The narrow spectral width also allows for high bit rates since it reduces the effect of chromatic dispersion. Furthermore, semiconductor lasers can be modulated directly at high frequencies because of short recombination time.

Commonly used classes of semiconductor laser transmitters used in fiber optics include VCSEL (Vertical-Cavity Surface-Emitting Laser), Fabry–Pérot and DFB (Distributed Feed Back).

Laser diodes are often directly modulated, that is the light output is controlled by a current applied directly to the device. For very high data rates or very long distance *links, a laser source may be operated* continuous wave, and the light modulated by an external device such as an electro-absorption modulator or Mach–Zehnder interferometer. External modulation increases the achievable link distance by eliminating laser chirp, which broadens the linewidth of directly modulated lasers, increasing the chromatic dispersion in the fiber.

A transceiver is a device combining a transmitter and a receiver in a single housing.

Receivers

The main component of an optical receiver is a photodetector, which converts light into electricity using the photoelectric effect. The primary photodetectors for telecommunications are made from Indium gallium arsenide The photodetector is typically a semiconductor-based photodiode. Several types of photodiodes include p-n photodiodes, p-i-n photodiodes, and avalanche photodiodes. Metal-semiconductor-metal (MSM) photodetectors are also used due to their suitability for circuit integration in regenerators and wavelength-division multiplexers.

Optical-electrical converters are typically coupled with a transimpedance amplifier and a limiting amplifier to produce a digital signal in the electrical domain from the incoming optical signal, which may be attenuated and distorted while passing through the channel. Further signal processing such as clock recovery from data (CDR) performed by a phase-locked loop may also be applied before the data is passed on.

Fiber Cable Types

An optical fiber cable consists of a core, cladding, and a buffer (a protective outer coating), in which the cladding guides the light along the core by using the method of total internal reflection. The core and the cladding (which has a lower-refractive-index) are usually made of high-quality silica glass, although they can both be made of plastic as well. Connecting two optical fibers is done by fusion splicing or mechanical splicing and requires special skills and interconnection technology due to the microscopic precision required to align the fiber cores.

A cable reel trailer with conduit that can carry optical fiber

Multi-mode optical fiber in an underground service pit

Two main types of optical fiber used in optic communications include multi-mode optical fibers and single-mode optical fibers. A multi-mode optical fiber has a larger core (≥ 50 micrometers), allowing less precise, cheaper transmitters and receivers to connect to it as well as cheaper connectors. However, a multi-mode fiber introduces multimode distortion, which often limits the bandwidth and length of the link. Furthermore, because of its higher dopant content, multi-mode fibers are usually expensive and exhibit higher attenuation. The core of a single-mode fiber is smaller (<10 micrometers) and requires more expensive components and interconnection methods, but allows much longer, higher-performance links.

In order to package fiber into a commercially viable product, it typically is protectively coated by using ultraviolet (UV), light-cured acrylate polymers, then terminated with optical fiber connectors, and finally assembled into a cable. After that, it can be laid in the ground and then run through the walls of a building and deployed aerially in a manner similar to copper cables. These fibers require less maintenance than common twisted pair wires, once they are deployed.

Specialized cables are used for long distance subsea data transmission, e.g. transatlantic communications cable. New (2011–2013) cables operated by commercial enterprises (Emerald Atlantis, Hibernia Atlantic) typically have four strands of fiber and cross the Atlantic (NYC-London) in 60-70ms. Cost of each such cable was about $300M in 2011. *source: The Chronicle Herald.*

Another common practice is to bundle many fiber optic strands within long-distance power transmission cable. This exploits power transmission rights of way effectively, ensures a power company can own and control the fiber required to monitor its own devices and lines, is effectively immune to tampering, and simplifies the deployment of smart grid technology.

Amplifier

The transmission distance of a fiber-optic communication system has traditionally been limited by fiber attenuation and by fiber distortion. By using opto-electronic repeaters, these problems have been eliminated. These repeaters convert the signal into an electrical signal, and then use a transmitter to send the signal again at a higher intensity than was received, thus counteracting the loss incurred in the previous segment. Because of the high complexity with modern wavelength-division multiplexed signals (including the fact that they had to be installed about once every 20 km), the cost of these repeaters is very high.

An alternative approach is to use an optical amplifier, which amplifies the optical signal directly without having to convert the signal into the electrical domain. It is made by doping a length of fiber with the rare-earth mineral erbium, and *pumping it with light from a* laser with a shorter wavelength than the communications signal (typically 980 nm). Amplifiers have largely replaced repeaters in new installations.

Wavelength-Division Multiplexing

Wavelength-division multiplexing (WDM) is the practice of multiplying the available capacity of optical fibers through use of parallel channels, each channel on a dedicated wavelength of light. This requires a wavelength division multiplexer in the transmitting equipment and a demultiplexer (essentially a spectrometer) in the receiving equipment. Arrayed waveguide gratings are commonly used for multiplexing and demultiplexing in WDM. Using WDM technology now commercially available, the bandwidth of a fiber can be divided into as many as 160 channels to support a combined bit rate in the range of 1.6 Tbit/s.

Parameters

Bandwidth–Distance Product

Because the effect of dispersion increases with the length of the fiber, a fiber transmission system is often characterized by its *bandwidth–distance product, usually expressed in units of* MHz·km. This value is a product of bandwidth and distance because there is a trade off between the bandwidth of the signal and the distance it can be carried. For example, a common multi-mode fiber with bandwidth–distance product of 500 MHz·km could carry a 500 MHz signal for 1 km or a 1000 MHz signal for 0.5 km.

Engineers are always looking at current limitations in order to improve fiber-optic communication, and several of these restrictions are currently being researched.

Record Speeds

Each fiber can carry many independent channels, each using a different wavelength of light (wavelength-division multiplexing). The net data rate (data rate without overhead bytes) per fiber is the per-channel data rate reduced by the FEC overhead, multiplied by the number of channels (usually up to eighty in commercial dense WDM systems as of 2008.

Year	Organization	Effective speed	WDM channels	Per channel speed	Distance
2009	Alcatel-Lucent	15 Tbit/s	155	100 Gbit/s	90 km
2010	NTT	69.1 Tbit/s	432	171 Gbit/s	240 km
2011	KIT	26 Tbit/s	1	26 Tbit/s	50 km
2011	NEC	101 Tbit/s	370	273 Gbit/s	165 km
2012	NEC, Corning	1.05 Petabit/s	12 core fiber		52.4 km

While the physical limitations of electrical cable prevent speeds in excess of 10 Gigabits per second, the physical limitations of fiber optics have not yet been reached.

In 2013, *New Scientist reported that a team at the* University of Southampton had achieved a throughput of 73.7 Tbit per second, with the signal traveling at 99.7% the speed of light through a hollow-core photonic crystal fiber.

Dispersion

For modern glass optical fiber, the maximum transmission distance is limited not by direct material absorption but by several types of dispersion, or spreading of optical pulses as they travel along the fiber. Dispersion in optical fibers is caused by a variety of factors. Intermodal dispersion, caused by the different axial speeds of different transverse modes, limits the performance of multi-mode fiber. Because single-mode fiber supports only one transverse mode, intermodal dispersion is eliminated.

In single-mode fiber performance is primarily limited by chromatic dispersion (also called group velocity dispersion), which occurs because the index of the glass varies slightly depending on the wavelength of the light, and light from real optical transmitters necessarily has nonzero spectral width (due to modulation). Polarization mode dispersion, another source of limitation, occurs because although the single-mode fiber can sustain only one transverse mode, it can carry this mode with two different polarizations, and slight imperfections or distortions in a fiber can alter the propagation velocities for the two polarizations. This phenomenon is called fiber birefringence and can be counteracted by polarization-maintaining optical fiber. Dispersion limits the bandwidth of the fiber because the spreading optical pulse limits the rate that pulses can follow one another on the fiber and still be distinguishable at the receiver.

Some dispersion, notably chromatic dispersion, can be removed by a 'dispersion compensator'. This works by using a specially prepared length of fiber that has the opposite dispersion to that induced by the transmission fiber, and this sharpens the pulse so that it can be correctly decoded by the electronics.

Attenuation

Fiber attenuation, which necessitates the use of amplification systems, is caused by a combination of material absorption, Rayleigh scattering, Mie scattering, and connection losses. Although material absorption for pure silica is only around 0.03 dB/km (modern fiber has attenuation around 0.3 dB/km), impurities in the original optical fibers caused attenuation of about 1000 dB/km. Other forms of attenuation are caused by physical stresses to the fiber, microscopic fluctuations in density, and imperfect splicing techniques.

Transmission Windows

Each effect that contributes to attenuation and dispersion depends on the optical wavelength. There are wavelength bands (or windows) where these effects are weakest, and these are the most favorable for transmission. These windows have been standardized, and the currently defined bands are the following:

Band	Description	Wavelength Range
O band	original	1260 to 1360 nm
E band	extended	1360 to 1460 nm
S band	short wavelengths	1460 to 1530 nm
C band	conventional ("erbium window")	1530 to 1565 nm
L band	long wavelengths	1565 to 1625 nm
U band	ultralong wavelengths	1625 to 1675 nm

Note that this table shows that current technology has managed to bridge the second and third windows that were originally disjoint.

Historically, there was a window used below the O band, called the first window, at 800-900 nm; however, losses are high in this region so this window is used primarily for short-distance communications. The current lower windows (O and E) around 1300 nm have much lower losses. This region has zero dispersion. The middle windows (S and C) around 1500 nm are the most widely used. This region has the lowest attenuation losses and achieves the longest range. It does have some dispersion, so dispersion compensator devices are used to remove this.

Regeneration

When a communications link must span a larger distance than existing fiber-optic technology is capable of, the signal must be *regenerated at intermediate points in the link by* optical communications repeaters. Repeaters add substantial cost to a communication system, and so system designers attempt to minimize their use.

Recent advances in fiber and optical communications technology have reduced signal degradation so far that *regeneration of the optical signal is only needed over distances of hundreds of kilometers. This has greatly reduced the cost of optical networking, particularly over undersea spans where the cost and reliability of repeaters is one of the key factors determining the performance of the whole cable system. The main advances contributing to these performance improvements are dispersion management, which seeks to balance the effects of dispersion against non-linearity; and* solitons, which use nonlinear effects in the fiber to enable dispersion-free propagation over long distances.

Last Mile

Although fiber-optic systems excel in high-bandwidth applications, optical fiber has been slow to achieve its goal of fiber to the premises or to solve the last mile problem. However, as bandwidth demand increases, more and more progress towards this goal can be observed. In Japan, for instance EPON has largely replaced DSL as a broadband Internet source. South Korea's KT also provides a service called FTTH (Fiber To The Home), which provides fiber-optic connections to the subscriber's home. The largest FTTH deployments are in Japan, South Korea, and China. Singapore started implementation of their all-fiber Next Generation Nationwide Broadband Network (Next Gen NBN), which is slated for completion in 2012 and is being installed by OpenNet. Since they began rolling out services in September 2010, Network coverage in Singapore has reached 85% nationwide.

In the US, Verizon Communications provides a FTTH service called FiOS to select high-ARPU (Average Revenue Per User) markets within its existing territory. The other major surviving ILEC (or Incumbent Local Exchange Carrier), AT&T, uses a FTTN (Fiber To The Node) service called U-verse with twisted-pair to the home. Their MSO competitors employ FTTN with coax using HFC. All of the major access networks use fiber for the bulk of the distance from the service provider's network to the customer.

Also in the US, Wilson Utilities, located in Wilson, North Carolina, has implemented FTTH and has successfully achieved 1 gigabit fiber to the home. This was implemented in late 2013. Wilson Utilities first rolled out their FTTH (Fiber to the Home) in 2012 with speeds offerings of 20/40/60/100 megabits per second. Their service is referred to as GreenLight.

Some other small cities in the US, such as Morristown, TN, have had their local utility company, Morristown Utility Systems in this case, deploy FTTH, offering symmetric gigabit speeds to each subscriber (though most are 50/50 or 100/100 mbit). It's called MUS Fibernet. AT&T and others have aggressively sought legislation at the state level to prevent further competition from municipalities, despite their low investment in rural areas.

The globally dominant access network technology is EPON (Ethernet Passive Optical Network). In Europe, and among telcos in the United States, BPON (ATM-based Broadband PON) and GPON (Gigabit PON) had roots in the FSAN (Full Service Access Network) and ITU-T standards organizations under their control.

Comparison with Electrical Transmission

The choice between optical fiber and electrical (or copper) transmission for a particular system is made based on a number of trade-offs. Optical fiber is generally chosen for systems requiring higher bandwidth or spanning longer distances than electrical cabling can accommodate.

A mobile fiber optic splice lab used to access and splice underground cables

The main benefits of fiber are its exceptionally low loss (allowing long distances between amplifiers/repeaters), its absence of ground currents and other parasite signal and power issues common to long parallel electric conductor runs (due to its reliance on light rather than electricity for transmission, and the dielectric nature of fiber optic), and its inherently high data-carrying capacity. Thousands of electrical links would be required to replace a single high bandwidth fiber cable. Another benefit of fibers is that even when run alongside each other for long distances, fiber cables experience effectively no crosstalk, in contrast to some types of electrical transmission lines. Fiber can be installed in areas with high electromagnetic interference (EMI), such as alongside utility lines, power lines, and railroad tracks. Nonmetallic all-dielectric cables are also ideal for areas of high lightning-strike incidence.

For comparison, while single-line, voice-grade copper systems longer than a couple of kilometers require in-line signal repeaters for satisfactory performance; it is not unusual for optical systems to go over 100 kilometers (62 mi), with no active or passive processing. Single-mode fiber cables are commonly available in 12 km lengths, minimizing the number of splices required over a long cable run. Multi-mode fiber is available in lengths up to 4 km, although industrial standards only mandate 2 km unbroken runs.

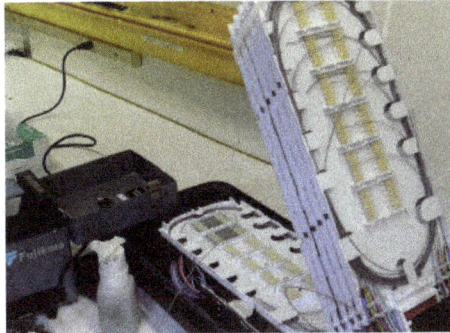

An underground fiber optic splice enclosure opened up

In short distance and relatively low bandwidth applications, electrical transmission is often preferred because of its

- Lower material cost, where large quantities are not required

- Lower cost of transmitters and receivers

- Capability to carry electrical power as well as signals (in appropriately designed cables)

- Ease of operating transducers in linear mode.

- Crosstalk from nearby cables and other parasitical unwanted signals increase profits from replacement and mitigation devices.

Optical fibers are more difficult and expensive to splice than electrical conductors. And at higher powers, optical fibers are susceptible to fiber fuse, resulting in catastrophic destruction of the fiber core and damage to transmission components.

Because of these benefits of electrical transmission, optical communication is not common in short box-to-box, backplane, or chip-to-chip applications; however, optical systems on those scales have been demonstrated in the laboratory.

In certain situations fiber may be used even for short distance or low bandwidth applications, due to other important features:

- Immunity to electromagnetic interference, including nuclear electromagnetic pulses.

- High electrical resistance, making it safe to use near high-voltage equipment or between areas with different earth potentials.

- Lighter weight—important, for example, in aircraft.

- No sparks—important in flammable or explosive gas environments.

- Not electromagnetically radiating, and difficult to tap without disrupting the signal—important in high-security environments.

- Much smaller cable size—important where pathway is limited, such as networking an existing building, where smaller channels can be drilled and space can be saved in existing cable ducts and trays.

- Resistance to corrosion due to non-metallic transmission medium

Optical fiber cables can be installed in buildings with the same equipment that is used to install copper and coaxial cables, with some modifications due to the small size and limited pull tension and bend radius of optical cables. Optical cables can typically be installed in duct systems in spans of 6000 meters or more depending on the duct's condition, layout of the duct system, and installation technique. Longer cables can be coiled at an intermediate point and pulled farther into the duct system as necessary.

Governing Standards

In order for various manufacturers to be able to develop components that function compatibly in fiber optic communication systems, a number of standards have been developed. The International Telecommunications Union publishes several standards related to the characteristics and performance of fibers themselves, including

- ITU-T G.651, "Characteristics of a 50/125 μm multimode graded index optical fibre cable"

- ITU-T G.652, "Characteristics of a single-mode optical fibre cable"

Other standards specify performance criteria for fiber, transmitters, and receivers to be used together in conforming systems. Some of these standards are:

- 100 Gigabit Ethernet

- 10 Gigabit Ethernet

- Fibre Channel

- Gigabit Ethernet

- HIPPI

- Synchronous Digital Hierarchy

- Synchronous Optical Networking

- Optical Transport Network (OTN)

TOSLINK is the most common format for digital audio cable using plastic optical fiber to connect digital sources to digital receivers.

Laser Pointer

A laser pointer or laser pen is a small handheld device with a power source (usually a battery) and a laser diode emitting a very narrow coherent low-powered laser beam of visible light, intended to be used to highlight something of interest by illuminating it with a small bright spot of colored light. Power is restricted in most jurisdictions not to exceed 5 mW.

The small width of the beam and low power of typical laser pointers make the beam itself invisible in a reasonably clean atmosphere, only showing a point of light when striking an opaque surface. Some higher-powered laser pointers project a visible beam via scattering from dust particles or water droplets along the beam path. Higher-power and higher-frequency green or blue lasers may produce a beam visible even in clean air because of Rayleigh scattering from air molecules, especially when viewed in moderately-to-dimly lit conditions. The intensity of such scattering increases when these beams are viewed from angles near the beam axis. Such pointers, particularly in the green-light output range, are used as astronomical-object pointers for teaching purposes.

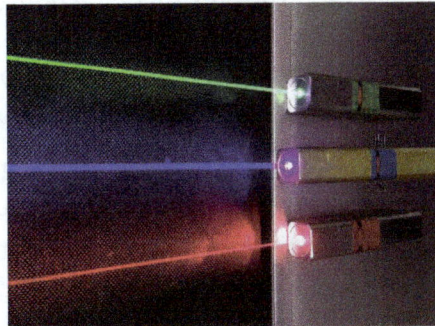

Red (635 nm), green (520 nm) and blue (445 nm) laser pointers

The low-cost availability of infrared (IR) diode laser modules of up to 1000 mW (1 watt) output has created a generation of IR-pumped, frequency doubled, green, blue, and violet diode-pumped solid-state laser pointers with visible power up to 300 mW. Because the invisible IR component in the beams of these visible lasers is difficult to filter out, and also because filtering it contributes extra heat which is difficult to dissipate in a small pocket "laser pointer" package, it is often left as a beam component in cheaper high-power pointers. This invisible IR component causes a degree of extra potential hazard in these devices when pointed at nearby objects and people.

Laser pointers make a potent signaling tool, even in daylight, and are able to produce a bright signal for potential search and rescue vehicles using an inexpensive, small and lightweight device of the type that could be routinely carried in an emergency kit.

If aimed at a person's eyes, laser pointers can cause temporary disturbances to vision. There is some evidence of rare minor permanent harm, but low-powered laser pointers are not seriously hazardous to health. They may be a major annoyance in some circumstances. A dot of light from a red laser pointer may be thought to be due to a laser gunsight. When pointed at aircraft at night, laser pointers may dazzle and distract pilots, and increasingly strict laws have been passed to ban this.

Colors and Wavelengths

Early laser pointers were helium–neon (HeNe) gas lasers and generated laser radiation at 633 nanometers (nm), usually designed to produce a laser beam with an output power under 1 milliwatt (mW). The least expensive laser pointers use a deep-red laser diode near the 650 nm wavelength. Slightly more expensive ones use a red-orange 635 nm diode, more easily visible because of the greater sensitivity of the human eye at 635 nm. Other colors are possible too, with the 532 nm green laser being the most common alternative. Yellow-orange laser pointers, at 593.5 nm, later became available. In September 2005 handheld blue laser pointers at 473 nm became available. In early 2010 "Blu-ray" (actually violet) laser pointers at 405 nm went on sale.

The apparent brightness of a spot from a laser beam depends on the optical power of the laser, the reflectivity of the surface, and the chromatic response of the human eye. For the same optical power, green laser light will seem brighter than other colors because the human eye is most sensitive at low light levels in the green region of the spectrum (wavelength 520–570 nm). Sensitivity decreases for longer (redder) and shorter (bluer) wavelengths.

The output power of a laser pointer is usually stated in milliwatts (mW). In the U.S. lasers are classified by the American National Standards Institute and Food and Drug Administration (FDA). Visible laser pointers (400–700 nm) operating at less than 1 mW power are Class 2 or II, and visible laser pointers operating with 1–5 mW power are Class 3A or IIIa. Class 3B or IIIb lasers generate between 5 and 500 mW; Class 4 or IV lasers generate more than 500 mW. The US FDA Code of Federal Regulations stipulates that "demonstration laser products" such as pointers must comply with applicable requirements for Class I, IIa, II, or IIIa devices.

Red and Red-Orange

These are the simplest pointers, as laser diodes are available in these wavelengths. The pointer is nothing more than a battery-powered laser diode. The first red laser pointers released in the early 1980s were large, unwieldy devices that sold for hundreds of dollars. Today, they are much smaller and generally cost very little. In the 21st century, diode-pumped solid-state (DPSS) red laser pointers emitting at 671 nm became available. Although this wavelength can be obtained directly with an inexpensive laser diode, higher beam quality and narrower spectral bandwidth are achieved through DPSS versions.

Yellow

Yellow laser pointers emitting at 593.5 nm became available in the last few years[*when?*]. Although they are based on the DPSS process, in this case two lasing lines of the ND:Y-VO4, 1064 nm and 1342 nm, are summed together with a nonlinear crystal. The complexity of this process makes these laser pointers inherently unstable and inefficient, with their outputs ranging from 1 mW to about 10 mW, greatly varying with temperature and usually mode-hopping if they get too hot or too cold. That is because such a complex process may require temperature stabilizers and active cooling, which can't be mounted into a small-sized host. Also, most smaller 593.5 nm pointers work in pulsed mode, so they can use smaller and less powerful pumping diodes. New 589 nm yellow laser pointers have been introduced using a more robust and secretive method[*clarify*] of harmonic generation from a DPSS laser system. This "sodium" wavelength, although only 4.5 nm away from the older 593.5 nm, appears more gold in colour compared to the more amber appearance of the 593.5 nm wavelength. Astronomical observatories use a specially tuned dye laser at 589.2 nm (yellow) to create a laser guide star for use with astronomical adaptive optics.

Green

Trails by a 15 mW green laser pointer in a time exposure of a living room at night

Green laser pointers appeared on the market around 2000 and are the most common type of DPSS lasers (also called *diode-pumped solid-state frequency-doubled, DPSSFD). They are more complicated than standard red laser pointers, because* laser diodes are not commonly available in this wavelength range. The green light is generated in an indirect process, beginning with a high-power (typically 100–300 mW) infrared aluminium gallium arsenide (AlGaAs) laser diode operating at 808 nm. The 808 nm light pumps a crystal of neodymium-doped yttrium orthovanadate (Nd:YVO4 or neodymium-doped yttrium aluminium garnet (Nd:YAG), or, less commonly, neodymium-doped yttrium lithium fluoride (Nd:YLF)), which lases deeper in the infrared at 1064 nm. This lasing action is due to an electronic transition in the fluorescent neodymium ion, Nd(III), which is present in all of these crystals.

The Nd:YVO4 or other Nd-doped crystal is coated on the diode side with a dielectric mirror that reflects at 808 nm and transmits at 1064 nm. The crystal is mounted on a copper block, acting as a heat sink; its 1064 nm output is fed into a crystal of potassium titanyl phosphate (KTP), mounted on a heat sink in the laser cavity resonator. The orientation of the crystals must be matched, as they are both anisotropic and the Nd:YVO4 outputs polarized light. This unit acts as a frequency doubler and halves the wavelength to the desired 532 nm. The resonant cavity is terminated by a dielectric mirror that reflects at 1064 nm and transmits at 532 nm. An infrared filter behind the mirror removes IR radiation from the output beam (this may be omitted or inadequate in less-expensive "pointer-style" green lasers), and the assembly ends in a collimator lens.

Nd:YVO4 is replacing other Nd-doped materials such as Nd:YAG and Nd:YLF in such systems because of lower dependency on the exact parameters of the pump diode (therefore allowing for higher tolerances), wider absorption band, lower lasing threshold, higher slope efficiency, linear polarization of output light, and single-mode output. For frequency doubling of higher-power lasers, lithium triborate (LBO) is used instead of KTP. Newer lasers use a composite Nd:YVO4/KTP crystal instead of two discrete ones.

Some green lasers operate in pulse or quasi-continuous wave (QCW) mode to reduce cooling problems and prolong battery life.

An announcement in 2009 of a direct green laser (which does not require doubling) promises much higher efficiencies and could foster the development of new color video projectors.

In 2012, Nichia and OSRAM developed and manufactured merchant high-power green laser diodes (515/520 nm), which can emit green laser directly.

Because even a low-powered green laser is visible at night through Rayleigh scattering from air molecules, this type of pointer is used by astronomers to easily point out stars and constellations. Green laser pointers can come in a variety of different output powers. The 5 mW green laser pointers (class IIIa) are the safest to use, and anything more powerful is usually not necessary for pointing purposes, since the beam is still visible in dark lighting conditions.

The United States Coast Guard requires their air crews to return to base if a green laser is pointed at them, and have their eyes examined for eye damage. People have been given up to five years in jail for aiming a green laser at an aircraft.

Blue

Blue laser pointers in specific wavelengths such as 473 nm usually have the same basic construction as DPSS green lasers. In 2006 many factories began production of blue laser modules for mass-storage devices, and these were used in laser pointers too. These were DPSS-type frequency-doubled devices. They most commonly emit a beam at 473 nm, which is produced by frequency doubling of 946 nm laser radiation from a diode-pumped Nd:YAG or Nd:YVO4 crystal (Nd-doped crystals usually produce a principal wavelength of 1064 nm, but with the proper reflective coating mirrors can be also made to lase at other "higher harmonic" non-principal neodymium wavelengths). For high output power, BBO crystals are used as frequency doublers; for lower powers, KTP is used. The Japanese company Nichia controlled 80% of the blue-laser-diode market in 2006.

Some vendors are now selling collimated diode blue laser pointers with measured powers exceeding 1,500 mW. However, since the claimed power of "laser pointer" products also includes the IR power (in DPSS technology only) still present in the beam (for reasons discussed below), comparisons on the basis of strictly visual-blue component from DPSS-type lasers remain problematic, and the information is often not available. Because of the higher neodymium harmonic used, and the lower efficiency of frequency-doubling conversion, the fraction of IR power converted to 473 nm blue laser light in optimally configured DPSS modules is typically 10–13%, about half that typical for green lasers (20–30%).

Blue lasers can also be fabricated directly with InGaN semiconductors, which produce blue light without frequency doubling. 450 nm (447 nm plus/minus 5 nm) blue laser diodes are currently available on the open market. Some blue diodes are capable of very high power; such as Nichia's NDB7K75 diode, which can continuously output over 5 watts of energy if overdriven. The devices are brighter for the same power than 405 nm violet laser diodes, since the longer wavelength is closer to the peak sensitivity of the human eye. Mass production of laser diodes for commercial devices like laser projectors have driven down prices. Recent popularity of the high-power version of these 447 nm pointers, which also have improved optics for better collimation and lower divergence, rivals the hazards associated with the use of these portable devices by persons of questionable intention and cost has diminished to be competitive with DPSS green lasers wavelengths.

Violet

Lasers emitting a violet light beam at 405 nm may be constructed with GaN (gallium nitride) semiconductors. This is close to ultraviolet, bordering on the very extreme of human vision, and can cause bright blue fluorescence, and thus a blue rather than violet spot, on many white surfaces, including white clothing, white paper, and projection screens, due to the widespread use of optical brighteners in the manufacture of products intended to appear brilliantly white. On ordinary non-fluorescent materials, and also on fog or dust, the color appears as a shade of deep violet that cannot be reproduced on monitors and print. A GaN laser emits 405 nm directly without a frequency doubler, eliminating the possibility of accidental dangerous infrared emission. These laser diodes are mass-produced for the reading and writing of data in Blu-ray drives (although the light emitted by the diodes is not blue, but distinctly violet). As of September 2011[update], 405 nm blue-violet laser diode modules with an optical power of 250 mW, based on GaN violet laser diodes made for Blu-ray disc readers, had reached the market from Chinese sources for prices of about US$60 including delivery.

At the same time, a few higher-powered (120 mW) 404–405 nm "violet" laser pointers have become available that are not based on GaN, but use DPSS frequency-doubler technology from 1-watt 808 nm GaAlAs infrared diode lasers. As with infrared-driven green laser pointers above, such devices are able to pop balloons and light matches, but this is as a result of an unfiltered high-power infrared component in the beam.

Applications

Pointing

A 5 mW green laser pointer directed at a palm tree at night. Note that the beam itself is visible through Rayleigh scattering.

Laser pointers are often used in educational and business presentations and visual demonstrations as an eye-catching pointing device. Laser pointers enhance verbal guidance given to students during surgery. The suggested mechanism of explanation is that the technology enables a more precise guidance of location and identification of anatomic structures.

Red laser pointers can be used in almost any indoor or low-light situation where pointing out details by hand may be inconvenient, such as in construction work or interior decorating. Green laser pointers can be used for similar purposes as well as outdoors in daylight or for longer distances.

Laser pointers are used in a wide range of applications. Green laser pointers can also be used for amateur astronomy. Green laser is visible at night due to Rayleigh scattering and airborne dust, allowing someone to point out individual stars to others nearby. Also, these green laser pointers are commonly used by astronomers worldwide at star parties or for conducting lectures in astronomy. Astronomy laser pointers are also commonly mounted on telescopes in order to align the telescope to a specific star or location. Laser alignment is much easier than aligning through using the eyepiece.

Industrial and Research Use

Laser pointers are used in industry. For instance, construction companies may use high quality laser pointers to enhance the accuracy of showing specific distances, while working on large-scale projects. They proved to be useful in this type of business because of their accuracy, which made them significant time-savers. What is essentially a laser pointer may be built into an infrared thermometer to identify where it is pointing, or be part of a laser level or other apparatus.

They may also be helpful in scientific research in fields such as photonics, chemistry, physics, and medicine.

Laser pointers are used in robotics, for example, for laser guidance to direct the robot to a goal position by means of a laser beam, i.e. showing goal positions to the robot optically instead of communicating them numerically. This intuitive interface simplifies directing the robot while visual feedback improves the positioning accuracy and allows for implicit localization.

Leisure and Entertainment

Entertainment is one of the other applications that have been found for lasers. The most common use of lasers in entertainment can be seen in special effects used in laser shows. Clubs, parties and outdoor concerts all use high-power lasers, with safety precautions, as a spectacle. Laser shows are often extravagant, using lenses, mirrors and smoke.

Lasers have also become a popular plaything for pets and other animals, namely cats and dogs, whose natural prey drive is triggered by the moving laser and will chase it and/or try to catch it as much as possible, but obviously never succeed. As a result, laser pointers have become a popular form of entertainment for many pet owners.

However, laser *pointers have few applications beyond actual pointing in the wider entertainment industry, and many venues ban entry to those in possession of pointers as a potential hazard. Very occasionally laser gloves, which are sometimes mistaken for pointers, are seen being worn by professional dancers on stage at shows. Unlike pointers, these usually produce low-power highly divergent beams to en-*

sure eye safety. Laser pointers have been used as props by magicians during magic shows.

As an example of the potential dangers of laser pointers brought in by audience members, at the Tomorrow Land Festival in Belgium in 2009, laser pointers brought in by members of the audience of 200 mW or greater were found to be the cause of eye damage suffered by several other members of the audience according to reports about the incident filed on the ILDA (International Laser Display Association's) Web site. The report says that the incident was investigated by several independent authorities, including the Belgium police, and that those authorities concluded that pointers brought in by the audience were the cause of the injuries.

Laser pointers can be used in hiking or outdoor activities. Higher-powered laser pointers are bright enough to scare away large wild animals which makes them useful for hiking and camping. In these circumstances a laser pointer can also serve as a handy survival tool, as it can be used as a rescue signal in emergencies which is visible to aircraft and other parties, during both day and night conditions, at extreme distances. For example, during the night in August 2010 two men and a boy were rescued from marshland after their red laser pen was spotted by rescue teams.

Weapons Systems

Accurately aligned laser pointers are used as laser gunsights to aim a firearm.

Some militaries use lasers to mark targets at night for aircraft. This is done to ensure that "friendly" and "enemy" targets are not mistaken. A friendly target may wear an IR emitting device that is only visible to those utilizing night vision (such as pilots.) To pinpoint the exact location of an enemy combatant, they would simply illuminate the target with a laser beam detectable by the attacking aircraft. This can be one of the most accurate ways of marking targets.

Hazards

Laser pointers can cause eye injuries directly, and dangerous situations by distraction. As of 2011[update], in spite of the very large number of pointers, and many incidents of malicious or careless use, no very serious and permanent eye injuries have been reported from low-power pointers, nor have any aviation or other accidents actually been reported to be caused by them, although the risk is clearly real.

Incorrect Power Rating

National Institute of Standards and Technology tests conducted on laser pointers labeled as Class IIIa or 3R in 2013 showed that about half of them emitted power at twice the Class limit, making their correct designation Class IIIb – more hazardous than Class IIIa. The highest measured power output was 66.5 milliwatts; more than

10 times the limit. Green laser light is generated from an infrared laser beam, which should be confined within the laser housing; however, more than 75% of the devices tested were found to emit infrared light in excess of the limit.

Malicious Use

Laser pointers, with their very long range, are often maliciously shone at people to distract or annoy them. This is considered particularly hazardous in the case of aircraft pilots, who may be dazzled or distracted at critical times. According to an MSNBC report there were over 2,836 incidents logged in the US by the FAA in 2010. Illumination by handheld green lasers is particularly serious, as the wavelength (532 nm) is near peak sensitivity of the dark-adapted eye and may appear to be 35 times brighter than a red laser of identical power output.

Irresponsible use of laser pointers is often frowned upon by members of the laser projector community who fear that their misuse may result in legislation affecting lasers designed to be placed within projectors and used within the entertainment industry. Others involved in activities where dazzling or distraction are dangerous are also a concern.

Another distressing and potentially dangerous misuse of laser pointers is to use them when the dot may reasonably be mistaken for that of a laser gunsight. Armed police have drawn their weapons in such circumstances.

Eye Injury

The output of laser pointers available to the general public is limited (and varies by country) in order to prevent accidental damage to the retina of human eyes. The U.K. Health Protection Agency recommended that "laser pointers generally available to the public should be restricted to less than 1 milliwatt as no injuries [like the one reported below to have caused retinal damage] have been reported at this power". In the U.S., regulatory authorities allow lasers up to 5 mW.

Studies have found that even low-power laser beams of not more than 5 mW can cause permanent retinal damage if gazed at for several seconds; however, the eye's blink reflex makes this highly unlikely. Such laser pointers have reportedly caused afterimages, flash blindness and glare, but not permanent damage, and are generally safe when used as intended.

A high-powered green laser pointer bought over the Internet was reported in 2010 to have caused a decrease of visual acuity from 6/6 to 6/12 (20/20 to 20/40); after two months acuity recovered to 6/6, but some retinal damage remained. The US FDA issued a warning after two anecdotal reports it received of eye injury from laser pointers.

Laser pointers available for purchase online can be capable of significantly higher power output than the pointers typically available in stores. Dubbed "Burning Lasers", these are designed to burn through light plastics and paper, and can have very similar external appearances to their low-power counterparts.

Studies in the early twenty-first century found that the risk to the human eye from accidental exposure to light from commercially available class IIIa laser pointers having powers up to 5 mW seemed rather small; however, prolonged viewing, such as deliberate staring into the beam for 10 or more seconds, can cause damage.

The UK Health Protection Agency warns against the higher-power typically green laser pointers available over the Internet, with power output of up to a few hundred milliwatts, as "extremely dangerous and not suitable for sale to the public."

Infrared Hazards of Diode-Pumped Solid-State Laser Pointers

Lasers classified as *pointers are intended to have outputs less than 5 mW total power* (Class 3R). At such power levels, an IR filter for a DPSS laser may not be required as the infrared (IR) output is relatively low and the brightness of the visible wavelength of the laser will cause the eye to react (blink reflex). However, higher-powered (> 5 mW) DPSS-type laser pointers have recently become available, usually through sources that do not follow laser safety regulations for laser packaging and labeling. These higher-powered lasers are often packaged in the same pointer-style housings as regular laser pointers, and usually lack the IR filters found in professional high-powered DPSS lasers, because of costs and additional efforts needed to accommodate them.

Though the IR from a DPSS laser is less collimated, the typical neodymium-doped crystals in such lasers do produce a true IR laser beam. The eye will usually react to the higher-powered visible light; however in higher power DPSS lasers the IR laser output can be significant. What poses a special hazard for this unfiltered IR output is its presence in conjunction with laser safety goggles designed to only block the visible wavelengths of the laser. Red goggles, for example, will block most green light from entering the eyes, but will pass IR light. The reduced light behind the goggles may also cause the pupils to dilate, increasing the hazard to the invisible IR light. Dual-frequency so-called YAG laser eyewear is significantly more expensive than single frequency laser eyewear, and is often not supplied with unfiltered DPSS pointer style lasers, which output 1064 nm IR laser light as well. These potentially hazardous lasers produce little or no visible beam when shone through the eyewear supplied with them, yet their IR-laser output can still be easily seen when viewed with an IR-sensitive video camera.

In addition to the safety hazards of unfiltered IR from DPSS lasers, the IR component may be inclusive of total output figures in some laser pointers.

Though green (532 nm) lasers are most common, IR filtering problems may also exist in other DPSS lasers, such as DPSS red (671 nm), yellow (589 nm) and blue (473 nm) lasers. These DPSS laser wavelengths are usually more exotic, more expensive, and generally manufactured with higher quality components, including filters, unless they are put into laser pointer style pocket-pen packages. Most red (635 nm, 660 nm), violet (405 nm) and darker blue (445 nm) lasers are generally built using dedicated laser diodes at the output frequency, not as DPSS lasers. These diode-based visible lasers do not produce IR light.

Regulations and Misuse

Since laser pointers became readily available, they have been misused, leading to the development of laws and regulations specifically addressing use of such lasers. Their very long range makes it difficult to find the source of a laser spot. In some circumstances they make people fear they are being targeted by weapons, as they are indistinguishable from dot type laser reticles. The very bright, small spot makes it possible to dazzle and distract drivers and aircraft pilots, and they can be dangerous to sight if aimed at the eyes.

In 1998, an audience member shone a laser at Kiss drummer Peter Criss's eyes while the band was performing Beth. After performing the song, Criss nearly stormed off the stage, and lead singer Paul Stanley challenged whoever shone the laser to fight him on stage:

In every crowd, there's one or two people who don't belong here, Now I know you want to bring it to school tomorrow when you go to sixth grade, but leave it at home when you go to the show.

— Paul Stanley,

In January 2005 a New Jersey man named David Banach was arrested for pointing a green laser pointer at a small jet flying overhead.

In football a laser pointer is a prohibited item at stadiums during FIFA tournaments and matches, according to FIFA Stadium Safety and Security Regulations, it is also prohibited in matches and competitions organised by UEFA. In 2008 laser pointers were aimed at players' eyes in a number of sport matches worldwide. Olympique Lyonnais was fined by UEFA because of a laser pointer beam aimed by a Lyon fan at Cristiano Ronaldo. In a World Cup final qualifier match held in Riyadh, Saudi Arabia between the home team and the South Korean team, South Korean goalkeeper Lee Woon-Jae was hit in the eye with a green laser beam. At the 2014 World Cup during the final group stage match between Algeria and Russia a green laser beam was directed on the face of Russian goalkeeper Igor Akinfeev. After the match the Algerian Football Federation was fined CHF50,000 (approx. £33,000/€41,100/US$56,200) by FIFA for the use of lasers and other violations of the rules by Algerian fans at the stadium.

In 2009 police in the United Kingdom began tracking the sources of lasers being shone at helicopters at night, logging the source using GPS, using thermal imaging cameras to see the suspect, and even the warm pointer if discarded, and calling in police dog teams. As of 2010 the penalty could be five years' imprisonment.

Despite legislation limiting the output of laser pointers in some countries, higher-power devices are currently produced in other regions (especially China and Hong Kong), and are frequently imported by customers who purchase them directly via Internet mail order. The legality of such transactions is not always clear; typically, the lasers are sold as research or OEM devices (which are not subject to the same power restrictions), with a disclaimer that they are not to be used as pointers. DIY videos are also often posted on Internet video sharing sites like YouTube which explain how to make a high-power laser pointer using the diode from an optical disc burner. As the popularity of these devices increased, manufacturers began manufacturing similar high-powered pointers. Warnings have been published on the dangers of such high-powered lasers. Despite the disclaimers, such lasers are frequently sold in packaging resembling that for laser pointers. Lasers of this type may not include safety features sometimes found on laser modules sold for research purposes.

There have been many incidents regarding, in particular, aircraft, and the authorities in many countries take them extremely serious. Many people have been convicted and sentenced, sometimes to several years' imprisonment.

Australia

In April 2008, citing a series of coordinated attacks on passenger jets in Sydney, the Australian government announced that it would restrict the sale and importation of certain laser items. The government had yet to determine which classes of laser pointers to ban. After some debate, the government voted to ban importation of lasers that emit a beam stronger than 1 mW, effective from July 1, 2008. Those whose professions require the use of a laser can apply for an exemption. In Victoria and the Australian Capital Territory a laser pointer with an accessible emission limit greater than 50 mW is classified as a prohibited weapon and any sale of such items must be recorded. In Western Australia, regulatory changes have classified laser pointers as controlled weapons and demonstration of a lawful reason for possession is required. The WA state government has also banned as of 2000 the manufacture, sale and possession of laser pointers higher than class 2. In New South Wales and the Australian Capital Territory the product safety standard for laser pointers prescribes that they must be a Class 1 or a Class 2 laser product. In February 2009 South African cricketer Wayne Parnell had a laser pointer directed at his eyes when attempting to take a catch, which he dropped. He denied that it was a reason for dropping the ball, but despite this the MCG decided to keep an eye out for the laser pointers. The laser pointer ban only applies to hand-held battery-powered laser devices and not laser modules.

In November 2015 a 14-year-old Tasmanian boy damaged both his eyes after shining a laser pen "… in his eyes for a very brief period of time". He burned his retinas near the macular, the area where most of a persons central vision is located. As a result, the boy has almost immediately lost 75% of his vision, with little hope of recovery.

Canada

New regulations controlling the importation and sale of laser pointers (portable, battery-powered) have been established in Canada in 2011 and are governed by Health Canada using the Consumer Protection Act for the prohibition of sale of Class 3B (IEC) or higher power lasers to "consumers" as defined in the Consumer Protection Act . Canadian federal regulation follows FDA (US Food & Drug Administration) CDRH, and IEC (International Electrotechnical Commission) hazard classification methods where manufacturers comply with the Radiation Emitting Devices Act. As of July 2011 three people had been charged under the federal Aeronautics Act, which carries a maximum penalty of $100,000 and five years in prison, for attempting to dazzle a pilot with a laser. Other charges that could be laid include mischief and assault.

Netherlands

Before 1998 Class 3A lasers were allowed. In 1998 it became illegal to trade Class 2 laser pointers that are "gadgets" (e.g. ball pens, key chains, business gifts, devices that will end up in children's possession, etc.). It is still allowed to trade Class 2 (< 1 mW) laser pointers proper, but they have to meet requirements regarding warnings and instructions for safe use in the manual. Trading of Class 3 and higher laser pointers is not allowed.

Sweden

The use of pointers with output power > 1 mW is regulated in public areas and school yards. From 1 January 2014 it is necessary to have a special permit in order to own a laser pointer with a classification of 3R, 3B or 4, i.e. over 1 mW.

United Kingdom

UK and most of Europe are now harmonized on Class 2 (<1 mW) for General presentation use laser pointers or laser pens. Anything above 1 mW is illegal for sale in the UK (import is unrestricted). Health and Safety regulation insists on use of Class 2 anywhere the public can come in contact with indoor laser light, and the DTI have urged Trading Standards authorities to use their existing powers under the General Product Safety Regulations 2005 to remove lasers above class 2 from the general market.

Since 2010, it is an offence in the UK to shine a light at an aircraft in flight so as to dazzle the pilot, whether intentionally or not, with a maximum penalty of a level 4 fine (currently £2500). It is also an offence to negligently or recklessly endanger an aircraft, with a maximum penalty of five years imprisonment and/or an unlimited fine.

To assist with enforcement, police helicopters use GPS and thermal imaging camera, together with dog teams on the ground, to help locate the offender; the discarded warm laser pointer is often visible on the thermal camera, and its wavelength can be matched to that recorded by an event recorder in the helicopter.

In 2014, a Flintshire 22-year-old was convicted of recklessly endangering the lives of a police search helicopter crew through the use of such a device, and awarded a five-month suspended sentence.

United States

Laser pointers are Class II or Class IIIa devices, with output beam power less than 5 milliwatts (<5 mW). According to U.S. Food and Drug Administration (FDA) regulations, more powerful lasers may not be sold or promoted as laser pointers. Also, any laser with class higher than IIIa (more than 5 milliwatts) requires a key-switch interlock and other safety features. Shining a laser pointer of any class at an aircraft is illegal and punishable by a fine of up to $11,000.

All laser products offered in commerce in the US must be registered with the FDA, regardless of output power.

In Utah it is a class C misdemeanor to point a laser pointer at a law enforcement officer and is an infraction to point a laser pointer at a moving vehicle.

In Arizona it is a Class 1 misdemeanor if a person "aims a laser pointer at a police officer if the person intentionally or knowingly directs the beam of light from an operating laser pointer at another person and the person knows or reasonably should know that the other person is a police officer." (Arizona Revised Statutes §13-1213)

On April 30, 2010, Clint Jason Brenner, 36, of Prescott, AZ was found guilty of two counts of endangerment, each a class 6 felony, and it was also found that each was a dangerous offense, for shining a handheld laser pointer at an Arizona Department of Public Safety helicopter in December 2009. He was given a sentence of two years in prison for each count, to run concurrently.

On November 2, 2009, Dana Christian Welch of Southern California was sentenced to 2.5 years in a federal prison after being found guilty of shining a hand-held laser light into the eyes of two pilots landing Boeing jets at John Wayne Airport.

Michigan

Public act 257 of 2003 makes it a felony for a person to "manufacture, deliver, possess, transport, place, use, or release" a "harmful electronic or electromagnetic device" for "an unlawful purpose"; also made into a felony is the act of causing "an individual to falsely believe that the individual has been exposed to a... harmful electronic or electromagnetic device."

Public act 328 of 1931 makes it a felony for a person to "sell, offer for sale, or possess" a "portable device or weapon from which an electric current, impulse, wave, or beam may be directed" and is designed "to incapacitate temporarily, injure, or kill".

Maine

Public law 264, H.P. 868 - L.D. 1271 criminalizes the knowing, intentional, and/or reckless use of an electronic weapon on another person, defining an electronic weapon as a portable device or weapon emitting an electric current, impulse, beam, or wave with disabling effects on a human being.

Massachusetts

Chapter 170 of the Acts of 2004, Section 140 of the General Laws, section 131J states: "No person shall possess a portable device or weapon from which an electric current, impulse, wave or beam may be directed, which current, impulse, wave or beam is designed to incapacitate temporarily, injure or kill, except ... Whoever violates this section shall be punished by a fine of not less than $500 nor more than $1,000 or by imprisonment in the house of correction for not less than 6 months nor more than 2 1/2 years, or by both such fine and imprisonment."

Magneto-Optical Trap

A magneto-optical trap (abbreviated MOT) is an apparatus that uses laser cooling with magneto-optical trapping in order to produce samples of cold, trapped, neutral atoms at temperatures as low as several microkelvins, two or three times the recoil limit. By combining the small momentum of a single photon with a velocity and spatially dependent absorption cross section and a large number of absorption-spontaneous emission cycles, atoms with initial velocities of hundreds of metres per second can be slowed to tens of centimetres per second.

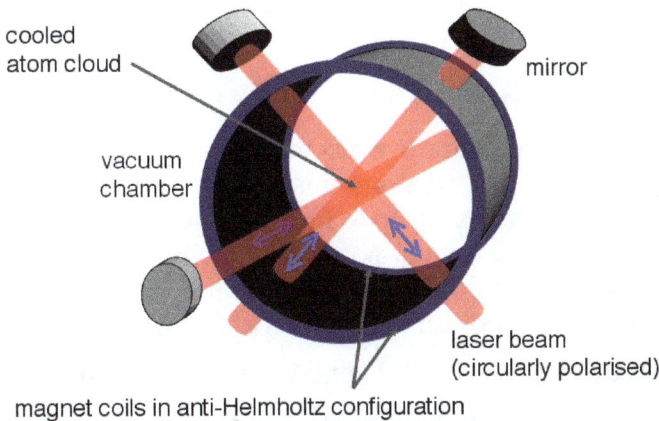

Experimental setup of the MOT

Although charged particles can be trapped using a Penning trap or a Paul trap using a combination of electric and magnetic fields, those traps do not work for neutral atoms.

Doppler Cooling

Photons have a momentum given by $\hbar k$ (where $\hbar$ is the reduced Planck constant and k the photon wavenumber), which is conserved in all atom-photon interactions. Thus, when an atom absorbs a photon, it is given a momentum kick in the direction of the photon before absorption. By detuning a laser beam to a frequency less than the resonant frequency (also known as red detuning), laser light is only absorbed if the light is frequency up-shifted by the Doppler effect, which occurs whenever the atom is moving towards the laser source. This applies a friction force to the atom whenever it moves towards a laser source.

For cooling to occur along all directions, the atom must see this friction force along all three Cartesian axes; this is most easily achieved by illuminating the atom with 3 orthogonal laser beams, which are then reflected back along the same direction.

Magnetic Trapping

Magnetic trapping is created by adding a spatially varying magnetic quadrupole field to the red detuned optical field needed for laser cooling. This causes a Zeeman shift in the magnetic-sensitive mf levels, which increases with the radial distance from the centre of the trap. Because of this, as an atom moves away from the centre of the trap, the atomic resonance is shifted closer to the frequency of the laser light, and the atom becomes more likely to get a photon kick towards the centre of the trap.

The direction of the kick is given by the polarisation of the light, which is either left or right handed circular, giving different interactions with the different mf levels. The correct polarisations are used so that photons moving towards the centre of the trap will be on resonance with the correct shifted atomic energy level, always driving the atom towards the centre.

Atomic Structure Necessary for Magneto-Optical Trapping

As a thermal atom at room temperature has many thousands of times the momentum of a single photon, the cooling of an atom must involve many absorption-spontaneous emission cycles, with the atom losing up to $\hbar k$ of momenta each cycle . Because of this, if an atom is to be laser cooled, it must possess a specific energy level structure known as a closed optical loop, where following an excitation-spontaneous emission event, the atom is always returned to its original state. 85Rubidium, for example, has a closed optical loop between the $5S_{1/2}$ $F = 3$ state and the $5P_{3/2}$ $F = 4$ state. Once in the excited state, the atom is forbidden from decaying to any of the $5P_{1/2}$ states, which would not conserve parity, and is also forbidden from decaying to the $5S_{1/2}$ $F = 2$ state, which would require an angular momentum change of -2, which cannot be supplied by a single photon.

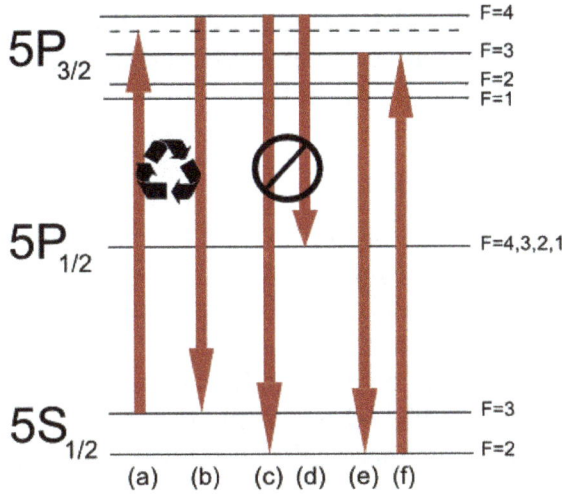

The lasers needed for the magneto-optical trapping of rubidium 85: (a) & (b) show the absorption (red detuned to the dotted line) and spontaneous emission cycle, (c) & (d) are forbidden transitions, (e) shows that if the cooling laser excites an atom to the $F = 3$ state, it is allowed to decay to the "dark" lower hyperfine, F=2 state, which would stop the cooling process, if it were not for the repumper laser (f).

Many atoms that do not contain closed optical loops can still be laser cooled, however, by using repump lasers which re-excite the population back into the optical loop after it has decayed to a state outside of the cooling cycle. The magneto-optical trapping of rubidium 85, for example, involves cycling on the closed $5S_{1/2}$ $F = 3 \rightarrow 5P_{3/2}$ $F = 4$ transition. On excitation, however, the detuning necessary for cooling gives a small, but non-zero overlap with the $5P_{3/2}$ $F = 3$ state. If an atom is excited to this state, which occurs roughly every thousand cycles, the atom is then free to decay either the $F = 3$, light coupled upper hyperfine state, or the $F = 2$ "dark" lower hyperfine state. If it falls back to the dark state, the atom stops cycling between ground and excited state, and the cooling and trapping of this atom stops. A repump laser which is resonant with the $5S_{1/2}$ $F = 2 \rightarrow 5P_{3/2}$ $F = 3$ transition is used to recycle the population back into the optical loop so that cooling can continue.

Apparatus

Laser

All magneto-optical traps require at least one trapping laser plus any necessary repumper lasers. These lasers need stability, rather than high power, requiring no more than the saturation intensity, but a linewidth much less than the Doppler width, usually several megahertz. Because of their low cost, compact size and ease of use, laser diodes are used for many of the standard MOT species while the linewidth and stability of these lasers is controlled using servo systems, which stabilises the lasers to an atomic frequency reference by using, for example, saturated absorption spectroscopy and the Pound-Drever-Hall technique to generate a locking signal.

By employing a 2-dimensional diffraction grating it is possible to generate the configuration of laser beams required for a magneto-optical trap from a single laser beam and thus have a very compact magneto-optical trap.

Vacuum Chamber

The MOT cloud is loaded from a background of thermal vapour, or from an atomic beam, usually slowed down to the capture velocity using a Zeeman slower. However, the trapping potential in a magneto-optical trap is small in comparison to thermal energies of atoms and most collisions between trapped atoms and the background gas supply enough energy to the trapped atom to kick it out of the trap. If the background pressure is too high, atoms are kicked out of the trap faster than they can be loaded, and the trap does not form. This means that the MOT cloud only forms in a vacuum chamber with a background pressure of less than 10 micropascals (10−10 bar).

The Limits to The Magneto-Optical Trap

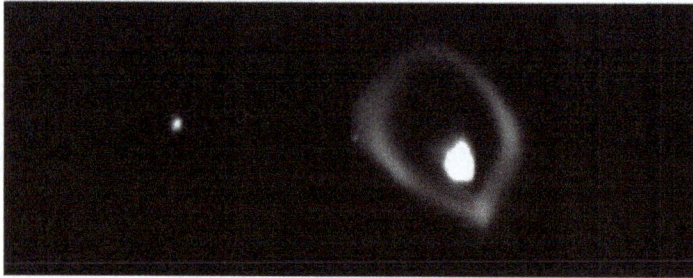

A MOT cloud in two different density regimes:If the density of the MOT is high enough, the MOT cloud goes from having a Gaussian density distribution (left), to something more exotic (right). In the right hand image, the density is so high that atoms have been blown out of the central trapping region by radiation pressure, to then form a toroidal racetrack mode around it.

The minimum temperature and maximum density of a cloud in a magneto-optical trap is limited by the spontaneously emitted photon in cooling each cycle. While the asymmetry in atom excitation gives cooling and trapping forces, the emission of the spontaneously emitted photon is in a random direction, and therefore contributes to a heating of the atom. Of the two $\hbar k$ kicks the atom receives in each cooling cycle, the first cools, and the second heats: a simple description of laser cooling which enables us to calculate a point at which these two effects reach equilibrium, and therefore define a lower temperature limit, known as the Doppler cooling limit.

The density is also limited by the spontaneously emitted photon. As the density of the cloud increases, the chance that the spontaneously emitted photon will leave the cloud without interacting with any further atoms tends to zero. The absorption, by a neighboring atom, of a spontaneously emitted photon gives a $2\hbar k$ momentum kick between the emitting and absorbing atom which can be seen as a repulsive force, similar to coulomb repulsion, which limits the maximum density of the cloud.

Application

As a result of low densities and speeds of atoms achieved by optical cooling, the mean free path in a ball of MOT cooled atoms is very long, and atoms may be treated as ballistic. This is useful for quantum information experiments where it is necessary to have long coherence times (the time an atom spends in a defined quantum state). Because of the continuous cycle of absorption and spontaneous emission, which causes decoherence, any quantum manipulation experiments must be performed with the MOT beams turned off. In this case, it is common to stop the expansion of the gases during quantum information experiments by loading the cooled atoms into a dipole trap.

A magneto-optical trap is usually the first step to achieving Bose–Einstein condensation. Atoms are cooled in a MOT down to a few times the recoil limit, and then evaporatively cooled which lowers the temperature and increases the density to the required phase space density.

A MOT of 133Cs was used to make some of the best measurements of CP violation.

Laser-Based Angle-Resolved Photoemission SpecTroscopy

Laser-based angle-resolved photoemission spectroscopy is a form of angle-resolved photoemission spectroscopy that uses a laser as the light source. Photoemission spectroscopy is a powerful and sensitive experimental technique to study surface physics. It is based on the photoelectric effect originally observed by Heinrich Hertz in 1887 and later explained by Albert Einstein in 1905 that when a material is shone by light, the electrons can absorb photons and escape from the material with the kinetic energy:

$E = hf - \phi$, where hf is the incident photon energy, ϕ the work function of the material. Since the kinetic energy of ejected electrons are highly associated with the internal electronic structure, by analyzing the photoelectron spectroscopy one can realize the fundamental physical and chemical properties of the material, such as the type and arrangement of local bonding, electronic structure and chemical composition.

In addition, because electrons with different momentum will escape from the sample in different directions, angle-resolved photoemission spectroscopy is widely used to provide the dispersive energy-momentum spectrum. The photoemission experiment is conducted using synchrotron radiation light source with typical photon energy of 20 – 100 eV. Synchrotron light is ideal for investigating two-dimensional surface systems and offers unparalleled flexibility to continuously vary the incident photon energy. However, due to the high costs to construct and maintain this accelerator, high competition for beam time, as well as the universal minimum electron mean free path in the material around the operating photon energy (20–100 eV) which leads to the fundamental hindrance to the three-dimensional bulk materials sensitivity, an alternative photon source for angle-resolved photoemission spectroscopy is desirable.

If femtosecond lasers are used, the method can easily be extended to access excited electronic states and electron dynamics by introducing a pump-probe scheme.

Laser-Based Arpes

Background

Table-top laser-based angle-resolved photoemission spectroscopy had been developed by some research groups. Daniel Dessau of University of Colorado, Boulder, made the first demonstration and applied this technique to explore superconducting system. The achievement not only greatly reduces the costs and size of facility, but also, most importantly, provides the unprecedented higher bulk sensitivity due to the low photon energy, typically 6 eV, and consequently the longer photoelectron mean free path (2–7 nm) in the sample. This advantage is extremely beneficial and powerful for the study of strongly correlated materials and high-Tc superconductors in which the physics of photoelectrons from the topmost layers might be different from the bulk. In addition to about one-order-of-magnitude improvement in the bulk sensitivity, the advance in the momentum resolution is also very significant: the photoelectrons will be more broadly dispersed in emission angle when the energy of incident photon decreases. In other words, for a given angular resolution of the electron spectrometer, the lower photon energy leads to higher momentum resolution. The typical momentum resolution of a 6 eV laser-based ARPES is approximately 8 times better than that of a 50 eV synchrotron radiation ARPES. Besides, the better momentum resolution due to low photon energy also results in less k-space accessible to ARPES which is helpful to the more precise spectrum analysis. For instance, in the 50 eV synchrotron ARPES, electrons from the first 4 Brillouin zones will be excited and scattered to contribute to the background of photoelectron analysis. However, the small momentum of 6 eV ARPES will only access some part of the first Brillouin zone and therefore only those electrons from small region of k-space can be ejected and detected as the background. The reduced inelastic scattering background is desirable while doing the measurement of weak physical quantities, in particular the high-Tc superconductors.

Experimental Realization

The first 6 eV laser-based ARPES system used a Kerr mode-locked Ti: sapphire oscillator is used and pumped with another frequency doubled Nd:Vanadate laser of 5 W and then generates 70 fs and 6 nJ pulses which are tunable around 840 nm (1.5 eV) with the 1 MHz repetition rate. Two stages of non-linear second harmonic generation of light are carried out through type I phase matching in β-barium borate and then the quadruple light with 210 nm (~ 6 eV) is generated and finally focused and directed into the ultra-high vacuum chamber as the low-energy photon source to investigate the electronic structure of the sample.

In the first demonstration, Dessau's group showed that the typical forth harmonic spectrum fits very well with the Gaussian profile with a full width at half maximum of 4.7 meV as well as presents a 200 μW power. The performance of high flux (~ 1014- 1015 photons/s) and narrow bandwidth makes the laser-based ARPES overwhelm the synchrotron radiation ARPES even though the best undulator beamlines are used. Another noticeable point is that one can make the quadruple light pass through either 1/4 wave plate or 1/2 wave plate which produces the circular polarization or any linear polarization light in the ARPES. Because the polarization of light can influence the signal to background ratio, the ability to control the polarization of light is a very significant improvement and advantage over the synchrotron ARPES. With the aforementioned favorable features, including lower costs for operating and maintenance, better energy and momentum resolution, and higher flux and ease of polarization control of photon source, the laser-based ARPES undoubtedly is an ideal candidate to be employed to conduct more sophisticated experiments in condensed matter physics.

Applications

High-Tc Superconductor

One way to show the powerful ability of laser-based ARPES is to study high Tc superconductors. The following figure references refer to this publication. Fig. 1 shows the experimental dispersion relation, binding energy vs. momentum, of the superconducting Bi2Sr2CaCu2O8+d along the nodal direction of the Brillouin zone. Fig. 1 (b) and Fig. 1 (c) are taken by the synchrotron light source of 28 eV and 52 eV, respectively, with the best undulator beamlines. The significantly sharper spectral peaks, the evidence of quasiparticles in the cuprate superconductor, by the powerful laser-based ARPES are shown in Fig. 1 (a). This is the first comparison of dispersive energy-momentum relation at low photon energy from table-top laser with higher energy from synchrotron ARPES. The much clearer dispersion in (a) indicates the improved energy-momentum resolution as well as many important physical features, such as overall band dispersion, Fermi surface, superconducting gaps, and a kink by electron-boson coupling, are successfully reproduced. It is foreseeable that in the near future the laser-based ARPES will be widely used to help condensed matter physicists get more detailed information about the nature of superconductivity in the exotic materials as well as other novel properties that cannot be observed by the state-of-the-art conventional experimental techniques.

Time-Resolved Electron Dynamics

Femtosecond laser-based ARPES can be extended to give spectroscopic access to excited states in time-resolved photoemission and two-photon photoelectron spectroscopy.

By pumping an electron to a higher level excited state with the first photon, the subsequent evolution and interactions of electronic states as a function of time can be studied by the second probing photon. The traditional pump-probe experiments usually measure the changes of some optical constants, which might be too complex to obtain the relevant physics. Since the ARPES can provide a lot of detailed information about the electronic structures and interactions, the pump-probe laser-based ARPES may study more complicated electronic systems with sub-picosecond resolution.

Summary and Perspective

Even though the angle-resolved synchrotron radiation source is widely used to investigate the surface dispersive energy-momentum spectrum, the laser-based ARPES can even provide more detailed and bulk-sensitive electronic structures with much better energy and momentum resolution, which are critically necessary for studying the strongly correlated electronic system, high-Tc superconductor, and phase transition in exotic quantum system. In addition, the lower costs for operating and higher photon flux make laser-based ARPES easier to be handled and more versatile and powerful among other modern experimental techniques for surface science.

References

- Taylor, Jim H.; Johnson, Mark R.; Crawford, Charles G. (2006). DVD Demystified. McGraw-Hill Professional. pp. 7–8. ISBN 0-07-142396-6.

- Edwin D. Reilly (2003). Milestones in Computer Science and Information Technology. Greenwood Press. ISBN 1-57356-521-0.

- Roy A. Allan (1 October 2001). A History of the Personal Computer: The People and the Technology. Allan Publishing. pp. 13–. ISBN 978-0-9689108-3-2.

- William E. Kasdorf (January 2003). The Columbia Guide to Digital Publishing. Columbia University Press. pp. 383–. ISBN 978-0-231-12499-7.

- Michael Shawn Malone (2007). Bill & Dave: How Hewlett and Packard Built the World's Greatest Company. Penguin. pp. 327–. ISBN 978-1-59184-152-4.

- Paul A. Strassmann (2008). The Computers Nobody Wanted: My Years with Xerox. Strassmann, Inc. pp. 126–. ISBN 978-1-4276-3270-8.

- S. Nagabhushana (2010). Lasers and Optical Instrumentation. I. K. International Pvt Ltd. pp. 269–. ISBN 978-93-80578-23-1.

- Roy Mayer (1999). Scientific Canadian: Invention and Innovation From Canada's National Research Council. Vancouver: Raincoast Books. ISBN 1-55192-266-5. OCLC 41347212.

- Roberto Scopigno; Susanna Bracci; Falletti, Franca; Mauro Matteini (2004). Exploring David. Diagnostic Tests and State of Conservation. Gruppo Editoriale Giunti. ISBN 88-09-03325-6.

- "Future Trends in Fiber Optics Communication" (PDF). WCE, London UK. July 2, 2014. ISBN 978-988-19252-7-5.

- Mary Kay Carson (2007). Alexander Graham Bell: Giving Voice To The World. Sterling Biographies. New York: Sterling Publishing. pp. 76–78. ISBN 978-1-4027-3230-0.

- Metzger, Robert M. (2012). The Physical Chemist's Toolbox. John Wiley & Sons. p. 207. ISBN 978-0-470-88925-1. Retrieved 15 June 2016.

- Metcalf, Harold J. & Straten, Peter van der (1999). Laser Cooling and Trapping. Springer-Verlag New York, Inc. ISBN 978-0-387-98728-6.

- Chapman, Francesca (1998-11-24). "Kiss Drummer Sees Red, Rips Dimwit With Laser Pointer". Philly.com. Retrieved 21 March 2015.

- Ross, Selina (5 November 2015). "Laser pointers not toys, optometrists warn, after Tasmanian teenager damages eyes". ABC News. Australian Broadcasting Corporation. Retrieved 5 November 2015.

- Gladwell, Malcolm (May 16, 2011). "Creation Myth - Xerox PARC, Apple, and the truth about innovation". The New Yorker. Retrieved 28 October 2013.

- He C, Morawska L, Taplin L (2012). "Particle emission characteristics of office printers. ;Environ Sci Technol. 2007] - PubMed - NCBI". ncbi.nlm.nih.gov. Retrieved 15 August 2012.

- "UK: Plane Bombs Explosions Were Possible Over U.S". Fox News. Archived from the original on March 29, 2012. Retrieved 2010-11-17.

- Murphy, Liam. "Case Study: Old Mine Workings". Subsurface Laser Scanning Case Studies. Liam Murphy. Retrieved 11 January 2012.

- "Green lasers halt Coast Guard air searches « Coast Guard Compass". Coastguard.dodlive.mil. 2011-02-28. Retrieved 2012-08-17.

Industrial Applications of Laser

The usage of lasers in industrial manufacturing is multifold and this chapter explores laser cooling, laser ablation, optical amplifier, optical tweezers, laser beam welding, laser beam machining and chirped pulse amplification. The various industries that use lasers include the military, aerospace and aeronautics, the security industry, the medical industry and many others. The chapter describes the type of lasers used and the application of these processes.

Laser Cooling

Laser cooling refers to a number of techniques in which atomic and molecular samples are cooled down to near absolute zero through the interaction with one or more laser fields. All laser cooling techniques rely on the fact that when an object (usually an atom) absorbs and re-emits a photon (a particle of light) its momentum changes. The temperature of an ensemble of particles is larger for larger variance in the velocity distribution of the particles. Laser cooling techniques combine atomic spectroscopy with the aforementioned mechanical effect of light to compress the velocity distribution of an ensemble of particles, thereby cooling the particles.

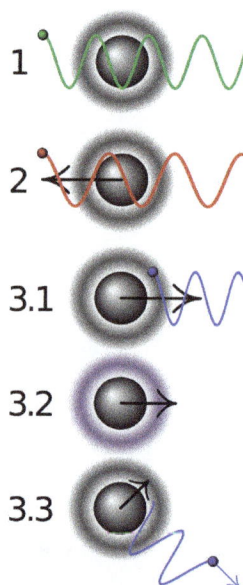

Simplified principle of doppler laser cooling:

1	A stationary atom sees the laser neither red- nor blue-shifted and does not absorb the photon.
2	An atom moving away from the laser sees it red-shifted and does not absorb the photon.
3.1	An atom moving towards the laser sees it blue-shifted and absorbs the photon, slowing the atom.
3.2	The photon excites the atom, moving an electron to a higher quantum state.
3.3	The atom re-emits a photon. As its direction is random, there is no net change in momentum over many absorption-emission cycles.

The first example of laser cooling, and also still the most common method (so much so that it is still often referred to simply as 'laser cooling') is doppler cooling. Other methods of laser cooling include:

- Sisyphus cooling

- Resolved sideband cooling

- Raman sideband cooling

- Velocity selective coherent population trapping (vscpt)

- Cavity mediated cooling

- Use of a zeeman slower

- Electromagnetically induced transparency (eit) cooling

Doppler Cooling

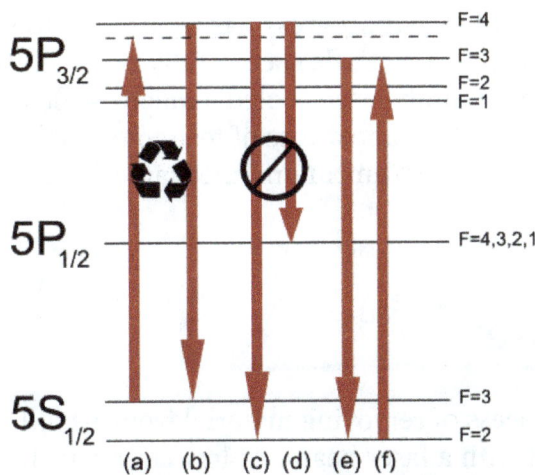

The lasers needed for the magneto-optical trapping of rubidium 85: (a) & (b) show the absorption (red detuned to the dotted line) and spontaneous emission cycle, (c) & (d) are forbidden transitions, (e) shows that if a cooling laser excites an atom to the $f=3$ state, it is allowed to decay to the "dark" lower hyperfine, $f=2$ state, which would stop the cooling process, if it were not for the repumper laser (f).

Doppler cooling, which is usually accompanied by a magnetic trapping force to give a magneto-optical trap, is by far the most common method of laser cooling. It is used to cool low density gases down to the doppler cooling limit, which for rubidium 85 is around 150 microkelvin.

In doppler cooling, the frequency of light is tuned slightly below an electronic transition in the atom. Because the light is detuned to the "red" (i.E., At lower frequency) of the transition, the atoms will absorb more photons if they move towards the light source, due to the doppler effect. Thus if one applies light from two opposite directions, the atoms will always scatter more photons from the laser beam pointing opposite to their direction of motion. In each scattering event the atom loses a momentum equal to the momentum of the photon. If the atom, which is now in the excited state, then emits a photon spontaneously, it will be kicked by the same amount of momentum, but in a random direction. Since the initial momentum loss was opposite to the direction of motion, while the subsequent momentum gain was in a random direction, the overall result of the absorption and emission process is to reduce the speed of the atom (provided its initial speed was larger than the recoil speed from scattering a single photon). If the absorption and emission are repeated many times, the average speed, and therefore the kinetic energy of the atom will be reduced. Since the temperature of a group of atoms is a measure of the average random internal kinetic energy, this is equivalent to cooling the atoms.

Uses

Laser cooling is primarily used to create ultracold atoms for experiments in quantum physics. These experiments are performed near absolute zero where unique quantum effects such as bose-einstein condensation can be observed. Laser cooling has primarily been used on atoms, but recent progress has been made toward laser cooling more complex systems. In 2010, a team at yale successfully laser-cooled a diatomic molecule. In 2007, an mit team successfully laser-cooled a macro-scale (1 gram) object to 0.8 K. In 2011, a team from the california institute of technology and the university of vienna became the first to laser-cool a (10 µm x 1 µm) mechanical object to its quantum ground state.

Laser Ablation

Laser ablation is the process of removing material from a solid (or occasionally liquid) surface by irradiating it with a laser beam. At low laser flux, the material is heated by the absorbed laser energy and evaporates or sublimates. At high laser flux, the material is typically converted to a plasma. Usually, laser ablation refers to removing material with a pulsed laser, but it is possible to ablate material with a continuous wave laser beam if the laser intensity is high enough.

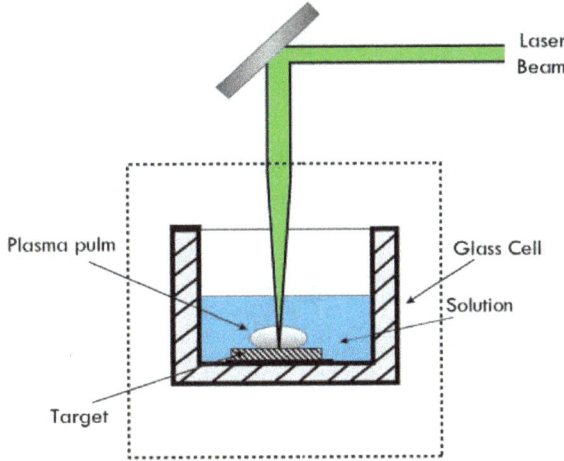

Preparation of nanoparticles by laser in solution

Fundamentals

The depth over which the laser energy is absorbed, and thus the amount of material removed by a single laser pulse, depends on the material's optical properties and the laser wavelength and pulse length. The total mass ablated from the target per laser pulse is usually referred to as ablation rate. Such features of laser radiation as laser beam scanning velocity and the covering of scanning lines can significantly influence the ablation process.

Laser pulses can vary over a very wide range of duration (milliseconds to femtoseconds) and fluxes, and can be precisely controlled. This makes laser ablation very valuable for both research and industrial applications.

Applications

The simplest application of laser ablation is to remove material from a solid surface in a controlled fashion. Laser machining and particularly laser drilling are examples; pulsed lasers can drill extremely small, deep holes through very hard materials. Very short laser pulses remove material so quickly that the surrounding material absorbs very little heat, so laser drilling can be done on delicate or heat-sensitive materials, including tooth enamel (laser dentistry). Several workers have employed laser ablation and gas condensation to produce nano particles of metal, metal oxides and metal carbides.

Also, laser energy can be selectively absorbed by coatings, particularly on metal, so CO_2 or nd:yag pulsed lasers can be used to clean surfaces, remove paint or coating, or prepare surfaces for painting without damaging the underlying surface. High power lasers clean a large spot with a single pulse. Lower power lasers use many small pulses which may be scanned across an area. The advantages are:

- No solvents are used, so it is environmentally friendly and operators are not exposed to chemicals. (Assuming nothing harmful is vaporized)

- It is relatively easy to automate, e.G., By using robots.

- The running costs are lower than dry media or dry-ice blasting, although the capital investment costs are much higher.

- The process is gentler than abrasive techniques, e.G. Carbon fibres within a composite material are not damaged.

- Heating of the target is minimal.

Another class of applications uses laser ablation to process the material removed into new forms either not possible or difficult to produce by other means. A recent example is the production of carbon nanotubes.

In march 1995 guo et al. Were the first to report the use of a laser to ablate a block of pure graphite, and later graphite mixed with catalytic metal. The catalytic metal can consist of elements such as cobalt, niobium, platinum, nickel, copper, or a binary combination thereof. The composite block is formed by making a paste of graphite powder, carbon cement, and the metal. The paste is next placed in a cylindrical mold and baked for several hours. After solidification, the graphite block is placed inside an oven with a laser pointed at it, and argon gas is pumped along the direction of the laser point. The oven temperature is approximately 1200 °c. As the laser ablates the target, carbon nanotubes form and are carried by the gas flow onto a cool copper collector. Like carbon nanotubes formed using the electric-arc discharge technique, carbon nanotube fibers are deposited in a haphazard and tangled fashion. Single-walled nanotubes are formed from the block of graphite and metal catalyst particles, whereas multi-walled nanotubes form from the pure graphite starting material.

A variation of this type of application is to use laser ablation to create coatings by ablating the coating material from a source and letting it deposit on the surface to be coated; this is a special type of physical vapor deposition called pulsed laser deposition (pld), and can create coatings from materials that cannot readily be evaporated any other way. This process is used to manufacture some types of high temperature superconductor.

Remote laser spectroscopy uses laser ablation to create a plasma from the surface material; the composition of the surface can be determined by analyzing the wavelengths of light emitted by the plasma.

Laser ablation is also used to create pattern, removing selectively coating from dichroic filter. This products are used in stage lighting for high dimensional projections, or for calibration of machine vision's instruments.

Propulsion

Finally, laser ablation can be used to transfer momentum to a surface, since the ablated material applies a pulse of high pressure to the surface underneath it as it expands. The effect is similar to hitting the surface with a hammer. This process is used in industry to work-harden metal surfaces, and is one damage mechanism for a laser weapon. It is also the basis of pulsed laser propulsion for spacecraft.

Manufacturing

The laser ablation of electronic semiconductors and microprocessors is now being pioneered in the uk to keep electronic manufacturers' designs confidential. The main reason is that it greatly reduces the risk of copying infringements.

Processes are currently being developed to use laser ablation in the removal of thermal barrier coating on high-pressure gas turbine components. Due to the low heat input, tbc removal can be completed with minimal damage to the underlying metallic coatings and parent material.

Science

Laser ablation is used in science to destroy nerves and other tissues to study their function.

For example, a species of pond snail, *helisoma trivolvis, can have their* sensory neurons laser ablated off when the snail is still an embryo to prevent use of those nerves.

Another example is the trochophore larva of *platynereis dumerilii, where the larval eye was ablated and the larvae was not phototactic, anymore.* However phototaxis in the nectochaete larva of *platynereis dumerilii is not mediated by the larval eyes, because the larva is still phototactic, even if the larval eyes are ablated. But if the adult eyes are ablated, then the nectochaete is not phototactic anymore and thus phototaxis in the nectochaete larva is mediated by the adult eyes.*

Laser ablation can also be used to destroy individual cells during embryogenesis of an organism, like *platynereis dumerilii, to study the effect of missing cells during development.*

Medicine

Laser ablation can be used on benign and malignant lesions in various organs, which is called laser-induced interstitial thermotherapy. The main applications currently involve the reduction of benign thyroid nodules and destruction of primary and secondary malignant liver lesions.

Laser ablation is also used to treat chronic venous insufficiency.

Optical Amplifier

Optical amplifiers are used to create laser guide stars.

An optical amplifier is a device that amplifies an optical signal directly, without the need to first convert it to an electrical signal. An optical amplifier may be thought of as a laser without an optical cavity, or one in which feedback from the cavity is suppressed. Optical amplifiers are important in optical communication and laser physics.

There are several different physical mechanisms that can be used to amplify a light signal, which correspond to the major types of optical amplifiers. In doped fibre amplifiers and bulk lasers, stimulated emission in the amplifier's gain medium causes amplification of incoming light. In semiconductor optical amplifiers (soas), electron-hole recombination occurs. In raman amplifiers, raman scattering of incoming light with phonons in the lattice of the gain medium produces photons coherent with the incoming photons. Parametric amplifiers use parametric amplification.

Laser Amplifiers

Almost any laser active gain medium can be pumped to produce gain for light at the wavelength of a laser made with the same material as its gain medium. Such amplifiers are commonly used to produce high power laser systems. Special types such as regenerative amplifiers and chirped-pulse amplifiers are used to amplify ultrashort pulses.

Doped Fibre Amplifiers

Doped fibre amplifiers (dfas) are optical amplifiers that use a doped optical fibre as a gain medium to amplify an optical signal. They are related to fibre lasers. The signal to be amplified and a pump laser are multiplexed into the doped fibre, and the signal is amplified through interaction with the doping ions. The most common example is the erbium doped fibre amplifier (edfa), where the core of a silica fibre is doped with trivalent erbium ions and can be efficiently pumped with a laser at a wavelength of 980 nm or 1,480 nm, and exhibits gain in the 1,550 nm region.

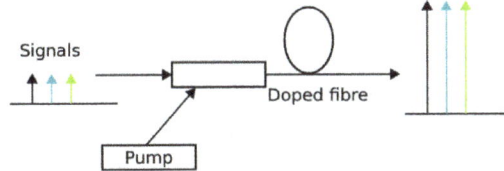

Signals

Doped fibre

Pump

Schematic diagram of a simple doped fibre amplifier

An *erbium-doped waveguide amplifier (edwa) is an optical amplifier that uses a waveguide to boost an optical signal.*

Amplification is achieved by stimulated emission of photons from dopant ions in the doped fibre. The pump laser excites ions into a higher energy from where they can decay via stimulated emission of a photon at the signal wavelength back to a lower energy level. The excited ions can also decay spontaneously (spontaneous emission) or even through nonradiative processes involving interactions with phonons of the glass matrix. These last two decay mechanisms compete with stimulated emission reducing the efficiency of light amplification.

The amplification window of an optical amplifier is the range of optical wavelengths for which the amplifier yields a usable gain. The amplification window is determined by the spectroscopic properties of the dopant ions, the glass structure of the optical fibre, and the wavelength and power of the pump laser.

Although the electronic transitions of an isolated ion are very well defined, broadening of the energy levels occurs when the ions are incorporated into the glass of the optical fibre and thus the amplification window is also broadened. This broadening is both homogeneous (all ions exhibit the same broadened spectrum) and inhomogeneous (different ions in different glass locations exhibit different spectra). Homogeneous broadening arises from the interactions with phonons of the glass, while inhomogeneous broadening is caused by differences in the glass sites where different ions are hosted. Different sites expose ions to different local electric fields, which shifts the energy levels via the stark effect. In addition, the stark effect also removes the degeneracy of energy states having the same total angular momentum (specified by the quantum number j). Thus, for example, the trivalent erbium ion (er+3) has a ground state with $j = 15/2$, and in the presence of an electric field splits into $j + 1/2 = 8$ sublevels with slightly different energies. The first excited state has $j = 13/2$ and therefore a stark manifold with 7 sublevels. Transitions from the $j = 13/2$ excited state to the $j = 15/2$ ground state are responsible for the gain at 1.5 Mm wavelength. The gain spectrum of the edfa has several peaks that are smeared by the above broadening mechanisms. The net result is a very broad spectrum (30 nm in silica, typically). The broad gain-bandwidth of fibre amplifiers make them particularly useful in wavelength-division multiplexed communications systems as a single amplifier can be utilized to amplify all signals being carried on a fibre and whose wavelengths fall within the gain window.

Basic Principle of Edfa

A relatively high-powered beam of light is mixed with the input signal using a wavelength selective coupler (wsc). The input signal and the excitation light must be at significantly different wavelengths. The mixed light is guided into a section of fibre with erbium ions included in the core. This high-powered light beam excites the erbium ions to their higher-energy state. When the photons belonging to the signal at a different wavelength from the pump light meet the excited erbium atoms, the erbium atoms give up some of their energy to the signal and return to their lower-energy state. A significant point is that the erbium gives up its energy in the form of additional photons which are exactly in the same phase and direction as the signal being amplified. So the signal is amplified along its direction of travel only. This is not unusual - when an atom "lases" it always gives up its energy in the same direction and phase as the incoming light. Thus all of the additional signal power is guided in the same fibre mode as the incoming signal. There is usually an isolator placed at the output to prevent reflections returning from the attached fibre. Such reflections disrupt amplifier operation and in the extreme case can cause the amplifier to become a laser. The erbium doped amplifier is a high gain amplifier.

Noise

The principal source of noise in dfas is amplified spontaneous emission (ase), which has a spectrum approximately the same as the gain spectrum of the amplifier. Noise figure in an ideal dfa is 3 db, while practical amplifiers can have noise figure as large as 6–8 db.

As well as decaying via stimulated emission, electrons in the upper energy level can also decay by spontaneous emission, which occurs at random, depending upon the glass structure and inversion level. Photons are emitted spontaneously in all directions, but a proportion of those will be emitted in a direction that falls within the numerical aperture of the fibre and are thus captured and guided by the fibre. Those photons captured may then interact with other dopant ions, and are thus amplified by stimulated emission. The initial spontaneous emission is therefore amplified in the same manner as the signals, hence the term *amplified spontaneous emission. Ase is emitted by the amplifier in both the forward and reverse directions, but only the forward ase is a direct concern to system performance since that noise will co-propagate with the signal to the receiver where it degrades system performance. Counter-propagating ase can, however, lead to degradation of the amplifier's performance since the ase can deplete the inversion level and thereby reduce the gain of the amplifier.*

Gain Saturation

Gain is achieved in a dfa due to population inversion of the dopant ions. The inversion level of a dfa is set, primarily, by the power of the pump wavelength and the power at the amplified wavelengths. As the signal power increases, or the pump power decreases, the inversion level will reduce and thereby the gain of the amplifier will be reduced. This effect is known as gain saturation – as the signal level increases, the amplifier saturates and cannot produce any more output power, and therefore the gain reduces. Saturation is also commonly known as gain compression.

To achieve optimum noise performance dfas are operated under a significant amount of gain compression (10 db typically), since that reduces the rate of spontaneous emission, thereby reducing ase. Another advantage of operating the dfa in the gain saturation region is that small fluctuations in the input signal power are reduced in the output amplified signal: smaller input signal powers experience larger (less saturated) gain, while larger input powers see less gain.

The leading edge of the pulse is amplified, until the saturation energy of the gain medium is reached. In some condition, the width (fwhm) of the pulse is reduced.

Inhomogeneous Broadening Effects

Due to the inhomogeneous portion of the linewidth broadening of the dopant ions, the gain spectrum has an inhomogeneous component and gain saturation occurs, to a small extent, in an inhomogeneous manner. This effect is known as *spectral hole burning because a high power signal at one wavelength can 'burn' a hole in the gain for wavelengths close to that signal by saturation of the inhomogeneously broadened ions. Spectral holes vary in width depending on the characteristics of the optical fibre in question and the power of the burning signal, but are typically less than 1 nm at the short wavelength end of the c-band, and a few nm at the long wavelength end of the c-band. The depth of the holes are very small, though, making it difficult to observe in practice.*

Polarization Effects

Although the dfa is essentially a polarization independent amplifier, a small proportion of the dopant ions interact preferentially with certain polarizations and a small dependence on the polarization of the input signal may occur (typically < 0.5 Db). This is called polarization dependent gain (pdg). The absorption and emission cross sections of the ions can be modeled as ellipsoids with the major axes aligned at random in all directions in different glass sites. The random distribution of the orientation of the ellipsoids in a glass produces a macroscopically isotropic medium, but a strong pump laser induces an anisotropic distribution by selectively exciting those ions that are more aligned with the optical field vector of the pump. Also, those excited ions aligned with the signal field produce more stimulated emission. The change in gain is thus dependent on the alignment of the polarizations of the pump and signal lasers – i.E. Whether the two lasers are interacting with the same sub-set of dopant ions or not. In an ideal doped fibre without birefringence, the pdg would be inconveniently large. Fortunately, in optical fibres small amounts of birefringence are always present and, furthermore, the fast and slow axes vary randomly along the fibre length. A typical dfa has several tens of meters, long enough to already show this randomness of the birefringence axes. These two combined effects (which in transmission fibres give rise to polarization mode dispersion) produce a misalignment of the relative polarizations of the signal and pump lasers along the fibre, thus tending to average out the pdg. The result is that pdg is very difficult to observe in a single amplifier (but is noticeable in links with several cascaded amplifiers).

Erbium-Doped Optical Fibre Amplifiers

The erbium-doped fibre amplifier (edfa) is the most deployed fibre amplifier as its amplification window coincides with the third transmission window of silica-based optical fibre.

Two bands have developed in the third transmission window – the conventional, or c-band, from approximately 1525 nm – 1565 nm, and the long, or l-band, from approximately 1570 nm to 1610 nm. Both of these bands can be amplified by edfas, but it is normal to use two different amplifiers, each optimized for one of the bands.

The principal difference between c- and l-band amplifiers is that a longer length of doped fibre is used in l-band amplifiers. The longer length of fibre allows a lower inversion level to be used, thereby giving at longer wavelengths (due to the band-structure of erbium in silica) while still providing a useful amount of gain.

Edfas have two commonly used pumping bands – 980 nm and 1480 nm. The 980 nm band has a higher absorption cross-section and is generally used where low-noise performance is required. The absorption band is relatively narrow and so wavelength stabilised laser sources are typically needed. The 1480 nm band has a lower, but broader, absorption cross-section and is generally used for higher power amplifiers. A combination of 980 nm and 1480 nm pumping is generally utilised in amplifiers.

The optical fibre amplifier was invented by h. J. Shaw and michel digonnet at stanford university, california, in the early 1980s. The edfa was first demonstrated several years later by a group including david n. Payne, r. Mears, i.M jauncey and l. Reekie, from the university of southampton in collaboration with a group from at&t bell laboratories, e. Desurvire, p. Becker, and j. Simpson and the italian company pirelli . The dual-stage optical amplifier which enabled dense wave division multiplexing (dwdm,) was invented by stephen b. Alexander at ciena corporation.

Doped Fibre Amplifiers for Other Wavelength Ranges

Thulium doped fibre amplifiers have been used in the s-band (1450–1490 nm) and praseodymium doped amplifiers in the 1300 nm region. However, those regions have not seen any significant commercial use so far and so those amplifiers have not been the subject of as much development as the edfa. However, ytterbium doped fibre lasers and amplifiers, operating near 1 micrometre wavelength, have many applications in industrial processing of materials, as these devices can be made with extremely high output power (tens of kilowatts).

Semiconductor Optical Amplifier

Semiconductor optical amplifiers (soas) are amplifiers which use a semiconductor to provide the gain medium. These amplifiers have a similar structure to fabry–pérot laser diodes but with anti-reflection design elements at the end faces. Recent designs include anti-reflective coatings and tilted wave guide and window regions which can reduce end face reflection to less than 0.001%. Since this creates a loss of power from the cavity which is greater than the gain, it prevents the amplifier from acting as a laser. Another type of soa consists of two regions. One part has a structure of a fabry-pérot laser diode and the other has a tapered geometry in order to reduce the power density on the output facet.

Semiconductor optical amplifiers are typically made from group iii-v compound semiconductors such as gaas/algaas, inp/ingaas, inp/ingaasp and inp/inalgaas, though any direct band gap semiconductors such as ii-vi could conceivably be used. Such amplifiers are often used in telecommunication systems in the form of fibre-pigtailed components, operating at signal wavelengths between 0.85 Mm and 1.6 Mm and generating gains of up to 30 db.

The semiconductor optical amplifier is of small size and electrically pumped. It can be potentially less expensive than the edfa and can be integrated with semiconductor lasers, modulators, etc. However, the performance is still not comparable with the edfa. The soa has higher noise, lower gain, moderate polarization dependence and high nonlinearity with fast transient time. The main advantage of soa is that all four types of nonlinear operations (cross gain modulation, cross phase modulation, wavelength conversion and four wave mixing) can be conducted. Furthermore, soa can be run with a low power laser. This originates from the short nanosecond or less upper state lifetime, so that the gain reacts rapidly to changes of pump or signal power and the changes of gain also cause phase changes which can distort the signals. This nonlinearity presents the most severe problem for optical communication applications. However it provides the possibility for gain in different wavelength regions from the edfa. "Linear optical amplifiers" using gain-clamping techniques have been developed.

High optical nonlinearity makes semiconductor amplifiers attractive for all optical signal processing like all-optical switching and wavelength conversion. There has been much research on semiconductor optical amplifiers as elements for optical signal processing, wavelength conversion, clock recovery, signal demultiplexing, and pattern recognition.

Vertical-Cavity Soa

A recent addition to the soa family is the vertical-cavity soa (vcsoa). These devices are similar in structure to, and share many features with, vertical-cavity surface-emitting lasers (vcsels). The major difference when comparing vcsoas and vcsels is the reduced mirror reflectivities used in the amplifier cavity. With vcsoas, reduced feedback is necessary to prevent the device from reaching lasing threshold. Due to the extremely short cavity length, and correspondingly thin gain medium, these devices exhibit very low single-pass gain (typically on the order of a few percent) and also a very large free spectral range (fsr). The small single-pass gain requires relatively high mirror reflectivities to boost the total signal gain. In addition to boosting the total signal gain, the use of the resonant cavity structure results in a very narrow gain bandwidth; coupled with the large fsr of the optical cavity, this effectively limits operation of the vcsoa to single-channel amplification. Thus, vcsoas can be seen as amplifying filters.

Given their vertical-cavity geometry, vcsoas are resonant cavity optical amplifiers that operate with the input/output signal entering/exiting normal to the wafer surface. In addition to their small size, the surface normal operation of vcsoas leads to a number of advantages, including low power consumption, low noise figure, polarization insensitive gain, and the ability to fabricate high fill factor two-dimensional arrays on a single semiconductor chip. These devices are still in the early stages of research, though promising preamplifier results have been demonstrated. Further extensions to vcsoa technology are the demonstration of wavelength tunable devices. These mems-tunable vertical-cavity soas utilize a microelectromechanical systems (mems) based tuning mechanism for wide and continuous tuning of the peak gain wavelength of the amplifier. Soas have a more rapid gain response, which is in the order of 1 to 100 ps.

Tapered Amplifiers

For high output power and broader wavelength range, tapered amplifiers are used. These amplifiers consist of a lateral single-mode section and a section with a tapered structure, where the laser light is amplified. The tapered structure leads to a reduction of the power density at the output facet.

Typical parameters:

- Wavelength range: 633 to 1480 nm

- Input power: 10 to 50 mw

- Output power: up to 3 watt

Raman Amplifier

In a raman amplifier, the signal is intensified by raman amplification. Unlike the edfa and soa the amplification effect is achieved by a nonlinear interaction between the signal and a pump laser within an optical fibre. There are two types of raman amplifier: distributed and lumped. A distributed raman amplifier is one in which the transmission fibre is utilised as the gain medium by multiplexing a pump wavelength with signal wavelength, while a lumped raman amplifier utilises a dedicated, shorter length of fibre to provide amplification. In the case of a lumped raman amplifier highly nonlinear fibre with a small core is utilised to increase the interaction between signal and pump wavelengths and thereby reduce the length of fibre required.

The pump light may be coupled into the transmission fibre in the same direction as the signal (co-directional pumping), in the opposite direction (contra-directional pumping) or both. Contra-directional pumping is more common as the transfer of noise from the pump to the signal is reduced.

The pump power required for raman amplification is higher than that required by the edfa, with in excess of 500 mw being required to achieve useful levels of gain in a distributed amplifier. Lumped amplifiers, where the pump light can be safely contained to avoid safety implications of high optical powers, may use over 1 w of optical power.

The principal advantage of raman amplification is its ability to provide distributed amplification within the transmission fibre, thereby increasing the length of spans between amplifier and regeneration sites. The amplification bandwidth of raman amplifiers is defined by the pump wavelengths utilised and so amplification can be provided over wider, and different, regions than may be possible with other amplifier types which rely on dopants and device design to define the amplification 'window'.

Raman amplifiers have some fundamental advantages. First, raman gain exists in every fiber, which provides a cost-effective means of upgrading from the terminal ends. Second, the gain is nonresonant, which means that gain is available over the entire transparency region of the fiber ranging from approximately 0.3 To 2µm. A third advantage of raman amplifiers is that the gain spectrum can be tailored by adjusting the pump wavelengths. For instance, multiple pump lines can be used to increase the optical bandwidth, and the pump distribution determines the gain flatness. Another advantage of raman amplification is that it is a relatively broad-band amplifier with a bandwidth > 5 thz, and the gain is reasonably flat over a wide wavelength range.

However, a number of challenges for raman amplifiers prevented their earlier adoption. First, compared to the edfas, raman amplifiers have relatively poor pumping efficiency at lower signal powers. Although a disadvantage, this lack of pump efficiency also makes gain clamping easier in raman amplifiers. Second, raman amplifiers require a longer gain fiber. However, this disadvantage can be mitigated by combining gain and the dispersion compensation in a single fiber. A third disadvantage of raman amplifiers is a fast response time, which gives rise to new sources of noise, as further discussed below. Finally, there are concerns of nonlinear penalty in the amplifier for the wdm signal channels.

Optical parametric Amplifier

An optical parametric amplifier allows the amplification of a weak signal-impulse in a noncentrosymmetric nonlinear medium (e.G. Beta barium borate (bbo)). In contrast to the previously mentioned amplifiers, which are mostly used in telecommunication environments, this type finds its main application in expanding the frequency tunability of ultrafast solid-state lasers (e.G. Ti:sapphire). By using a noncollinear interaction geometry optical parametric amplifiers are capable of extremely broad amplification bandwidths.

Recent Achievements

The adoption of high power fiber lasers as an industrial material processing tool has been ongoing for several years and is now expanding into other markets including the medical and scientific markets. One key enhancement enabling penetration into the scientific market has been the improvements in high finesse fiber amplifiers, which are now capable of delivering single frequency linewidths (<5 khz) together with excellent beam quality and stable linearly polarized output. Systems meeting these specifications, have steadily progressed in the last few years from a few watts of output power, initially to the 10s of watts and now into the 100s of watts power level. This power scaling has been achieved with developments in the fiber technology, such as the adoption of stimulated brillouin scattering (sbs) suppression/mitigation techniques within the fiber, along with improvements in the overall amplifier design. The latest generation of high finesse, high power fiber amplifiers now deliver power levels exceeding what is available from commercial solid-state single frequency sources and are opening up new scientific applications as a result of the higher power levels and stable optimized performance.

Implementations

There are several simulation tools that can be used to design optical amplifiers. Popular commercial tools have been developed by optiwave systems and vpi systems.

Optical Tweezers

Optical tweezers (originally called "single-beam gradient force trap") are scientific instruments that use a highly focused laser beam to provide an attractive or repulsive force (typically on the order of piconewtons), depending on the refractive index mismatch to physically hold and move microscopic dielectric objects similar to tweezers.

Optical tweezers have been particularly successful in studying a variety of biological systems in recent years.

History and Development

The detection of optical scattering and gradient forces on micron sized particles was first reported in 1970 by arthur ashkin, a scientist working at bell labs. Years later, ashkin and colleagues reported the first observation of what is now commonly referred to as an optical tweezer: a tightly focused beam of light capable of holding microscopic particles stable in three dimensions.

One of the authors of this seminal 1986 paper, former united states secretary of energy steven chu, would go on to use optical tweezing in his work on cooling and trapping neutral atoms. This research earned chu the 1997 nobel prize in physics along with claude cohen-tannoudji and william d. Phillips. In an interview, steven chu described how ashkin had first envisioned optical tweezing as a method for trapping atoms. Ashkin was able to trap larger particles (10 to 10,000 nanometers in diameter) but it fell to chu to extend these techniques to the trapping of neutral atoms (0.1 Nanometers in diameter) utilizing resonant laser light and a magnetic gradient trap (cf. Magneto-optical trap).

In the late 1980s, arthur ashkin and joseph m. Dziedzic demonstrated the first application of the technology to the biological sciences, using it to trap an individual tobacco mosaic virus and *escherichia coli bacterium*. Throughout the 1990s and afterwards, researchers like carlos bustamante, james spudich, and steven block pioneered the use of optical trap force spectroscopy to characterize molecular-scale biological motors. These molecular motors are ubiquitous in biology, and are responsible for locomotion and mechanical action within the cell. Optical traps allowed these biophysicists to observe the forces and dynamics of nanoscale motors at the single-molecule level; optical trap force-spectroscopy has since led to greater understanding of the stochastic nature of these force-generating molecules.

Optical tweezers have proven useful in other areas of biology as well. For instance, in 2003 the techniques of optical tweezers were applied in the field of cell sorting; by creating a large optical intensity pattern over the sample area, cells can be sorted by their intrinsic optical characteristics. Optical tweezers have also been used to probe the cytoskeleton, measure the visco-elastic properties of biopolymers, and study cell motility. A bio-molecular assay in which clusters of ligand coated nano-particles are both optically trapped and optically detected after target molecule induced clustering was proposed in 2011 and experimentally demonstrated in 2013.

The kapitsa–dirac effect effectively demonstrated during 2001 uses standing waves of light to affect a beam of particles.

Researchers have also worked to convert optical tweezers from large, complex instruments to smaller, simpler ones, for use by those with smaller research budgets.

Physics

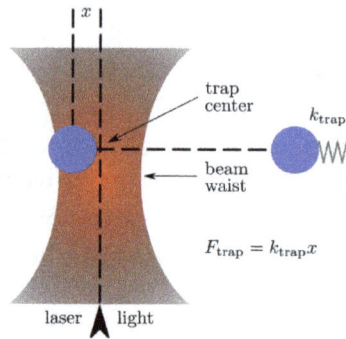

Dielectric objects are attracted to the center of the beam, slightly above the beam waist, as described in the text. The force applied on the object depends linearly on its displacement from the trap center just as with a simple spring system.

General Description

Optical tweezers are capable of manipulating nanometer and micron-sized dielectric particles by exerting extremely small forces via a highly focused laser beam. The beam is typically focused by sending it through a microscope objective. The narrowest point of the focused beam, known as the beam waist, contains a very strong electric field gradient. Dielectric particles are attracted along the gradient to the region of strongest electric field, which is the center of the beam. The laser light also tends to apply a force on particles in the beam along the direction of beam propagation. This is due to conservation of momentum: photons that are absorbed or scattered by the tiny dielectric particle impart momentum to the dielectric particle. This is known as the scattering force and results in the particle being displaced slightly downstream from the exact position of the beam waist, as seen in the figure.

Optical traps are very sensitive instruments and are capable of the manipulation and detection of sub-nanometer displacements for sub-micron dielectric particles. For this reason, they are often used to manipulate and study single molecules by interacting with a bead that has been attached to that molecule. Dna and the proteins and enzymes that interact with it are commonly studied in this way.

For quantitative scientific measurements, most optical traps are operated in such a way that the dielectric particle rarely moves far from the trap center. The reason for this is that the force applied to the particle is linear with respect to its displacement from the center of the trap as long as the displacement is small. In this way, an optical trap can be compared to a simple spring, which follows hooke's law.

Detailed View

Proper explanation of optical trapping behavior depends upon the size of the trapped particle relative to the wavelength of light used to trap it. In cases where the dimensions of the particle are much greater than the wavelength, a simple ray optics treatment is sufficient. If the wavelength of light far exceeds the particle dimensions, the particles can be treated as electric dipoles in an electric field. For optical trapping of dielectric objects of dimensions within an order of magnitude of the trapping beam wavelength, the only accurate models involve the treatment of either time dependent or time harmonic maxwell equations using appropriate boundary conditions.

Ray Optics

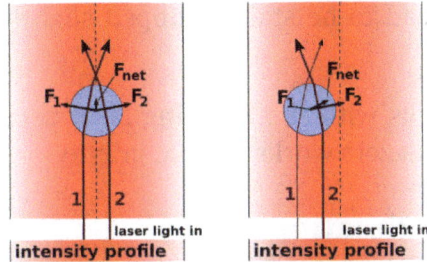

Ray optics explanation (unfocused laser). When the bead is displaced from the beam center (right image), the larger momentum change of the more intense rays cause a net force to be applied back toward the center of the laser. When the bead is laterally centered on the beam (left image), the resulting lateral force is zero. But an unfocused laser still causes a force pointing away from the laser.

In cases where the diameter of a trapped particle is significantly greater than the wavelength of light, the trapping phenomenon can be explained using ray optics. As shown in the figure, individual rays of light emitted from the laser will be refracted as it enters and exits the dielectric bead. As a result, the ray will exit in a direction different from which it originated. Since light has a momentum associated with it, this change in direction indicates that its momentum has changed. Due to newton's third law, there should be an equal and opposite momentum change on the particle.

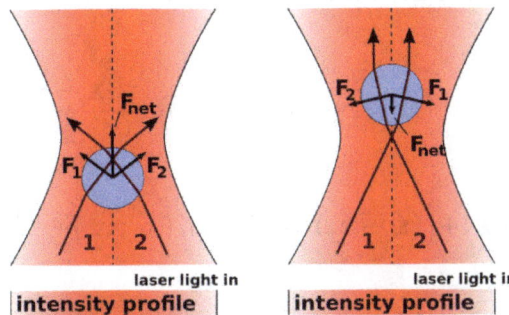

Ray optics explanation (focused laser). In addition to keeping the bead in the center of the laser, a focused laser also keeps the bead in a fixed axial position: the momentum change of the focused rays causes a force towards the laser focus, both when the bead is in front (left image) or behind (right image) the laser focus. So, bead will stay slightly behind the focus, where this force compensates the scattering force.

Most optical traps operate with a gaussian beam (temoo mode) profile intensity. In this case, if the particle is displaced from the center of the beam, as in the right part of the figure, the particle has a net force returning it to the center of the trap because more intense beams impart a larger momentum change towards the center of the trap than less intense beams, which impart a smaller momentum change away from the trap center. The net momentum change, or force, returns the particle to the trap center.

If the particle is located at the center of the beam, then individual rays of light are refracting through the particle symmetrically, resulting in no net lateral force. The net force in this case is along the axial direction of the trap, which cancels out the scattering force of the laser light. The cancellation of this axial gradient force with the scattering force is what causes the bead to be stably trapped slightly downstream of the beam waist.

The standard tweezers works with the trapping laser propagated in the direction of gravity and the inverted tweezers works against gravity.

Electric Dipole Approximation

In cases where the diameter of a trapped particle is significantly smaller than the wavelength of light, the conditions for rayleigh scattering are satisfied and the particle can be treated as a point dipole in an inhomogeneous electromagnetic field. The force applied on a single charge in an electromagnetic field is known as the lorentz force,

$$F_1 = q\left(E_1 + \frac{dx_1}{dt} \times B\right)$$

The force on the dipole can be calculated by substituting two terms for the electric field in the equation above, one for each charge. The polarization of a dipole is $p = qd$ where d is the distance between the two charges. For a point dipole, the distance is infinitesimal, $x_1 - x_2$ taking into account that the two charges have opposite signs, the force takes the form

$$F = q\left(E_1(x,y,z) - E_2(x,y,z) + \frac{d(x_1-x_2)}{dt} \times B\right)$$

$$= q\left(E_1(x,y,z) + ((x_1-x_2)\cdot\nabla)E - E_1(x,y,z) + \frac{d(x_1-x_2)}{dt} \times B\right)$$

NOTICE THAT THE E_1 CANCEL OUT. MULTIPLYING THROUGH BY THE CHARGE, q, CONVERTS POSITION, x, INTO POLARIZATION, p,

$$F = (p\cdot\nabla)E + \frac{dp}{dt} \times B$$

$$= \alpha \left[(\mathbf{E} \cdot \nabla) \mathbf{E} + \frac{d\mathbf{E}}{dt} \times \mathbf{B} \right],$$

Where in the second equality, it has been assumed that the dielectric particle is linear (i.E. $\mathbf{p} = \alpha \mathbf{E}$).

In the final steps, two equalities will be used: (1) a vector analysis equality, (2) one of maxwell's equations.

$$(\mathbf{E} \cdot \nabla) \mathbf{E} = \nabla \left(\frac{1}{2} E^2 \right) - \mathbf{E} \times (\nabla \times \mathbf{E})$$

$$\nabla \times \mathbf{E} = -\frac{\partial \mathbf{B}}{\partial t}$$

First, the vector equality will be inserted for the first term in the force equation above. Maxwell's equation will be substituted in for the second term in the vector equality. Then the two terms which contain time derivatives can be combined into a single term.

$$\mathbf{F} = \alpha \left[\frac{1}{2} \nabla E^2 - \mathbf{E} \times (\nabla \times \mathbf{E}) + \frac{d\mathbf{E}}{dt} \times \mathbf{B} \right]$$

$$= \alpha \left[\frac{1}{2} \nabla E^2 - \mathbf{E} \times \left(-\frac{d\mathbf{B}}{dt} \right) + \frac{d\mathbf{E}}{dt} \times \mathbf{B} \right]$$

$$= \alpha \left[\frac{1}{2} \nabla E^2 + \frac{d}{dt} (\mathbf{E} \times \mathbf{B}) \right].$$

The second term in the last equality is the time derivative of a quantity that is related through a multiplicative constant to the poynting vector, which describes the power per unit area passing through a surface. Since the power of the laser is constant when sampling over frequencies much shorter than the frequency of the laser's light ~1014 hz, the derivative of this term averages to zero and the force can be written as

$$\mathbf{F} = \frac{1}{2} \alpha \nabla E^2 = \frac{2\pi n_0 a^3}{c} \left(\frac{m^2 - 1}{m^2 + 2} \right) \nabla I(\mathbf{r}),$$

Where in the second part we have included the induced dipole of a spherical dielectric particle: $\alpha = 4\pi n_0^2 \epsilon_0 a^3 (m^2 - 1) / (m^2 + 2)$, where a is the particle radius, n_0 is the index of the medium and $m = n_1 / n_0$ is the relative index of the particle. The square of the magnitude of the electric field is equal to the intensity of the beam as a function of position. Therefore, the result indicates that the force on the dielectric particle, when treated as a point dipole, is proportional to the gradient along the intensity of the beam. In other words, the gradient force described here tends to attract the particle to the region of highest intensity. In reality, the scattering force of the light works against the gradient force in the axial direction of the trap, resulting in an equilibrium position that is displaced slightly downstream of the intensity maximum. Under the rayleigh approximation, we can also write the scattering force as

$$\mathbf{F}_{\text{scat}}(\mathbf{r}) = \frac{k^4\alpha^2}{6\pi c n_0^3 \epsilon_0^2} I(\mathbf{r})\hat{z} = \frac{8\pi n_0 k^4 a^6}{3c}\left(\frac{m^2-1}{m^2+2}\right)^2 I(\mathbf{r})\hat{z}.$$

Since the scattering is isotropic, the net momentum is transferred in the forward direction. On the quantum level, we picture this as incident photons all traveling in the forward direction and being scattered isotropically. By conservation of momentum, the sphere must accumulate the photons' original momenta, causing a forward force.

Harmonic Potential Approximation

A useful way to study the interaction of an atom in a gaussian beam is to look at the harmonic potential approximation of the intensity profile the atom experiences. In the case of the two-level atom, the potential experienced is related to its ac stark shift,

$$\Delta\mathbf{E}_{\text{AC Stark}} = \frac{3\pi c^2 \Gamma \mu}{2\omega_0^3 \delta} \mathbf{I(r,z)}$$

Where Γ is the natural line width of the excited state, μ is the electric dipole coupling, ω_o is the frequency of the transition, and δ is the defining or difference between the laser frequency and the transition frequency.

The intensity of a gaussian beam profile is characterized by the wavelength (λ), minimum waist (w_o), and power of the beam (P_o). The following formulas define the beam profile:

$$I(r,z) = I_0\left(\frac{w_0}{w(z)}\right)^2 e^{-\frac{2r^2}{w^2(z)}}$$

$$w(z) = w_0\sqrt{1+\left(\frac{z}{z_R}\right)^2}$$

$$z_R = \frac{\pi w_0^2}{\lambda}$$

$$P_0 = \frac{1}{2} \pi I_0 w_0^2$$

To approximate this gaussian potential in both the radial and axial directions of the beam, the intensity profile must be expanded to second order in z and r for $r = 0$ and $z = 0$ respectively and equated to the harmonic potential $\frac{1}{2} m(\omega_z^2 z^2 + \omega_r^2 r^2)$. These expansions are evaluated assuming fixed pow er.

$$\frac{\partial^2 I}{\partial z^2}\Big|_{r=0} = \frac{4P_0^2 \lambda^2}{\pi^3 w_0^6} z^2 = \frac{1}{2} m\omega_z^2 z^2$$

$$\frac{\partial^2 I}{\partial r^2}\Big|_{z=0} = \frac{4P_0^2}{\pi w_0^4} r^2 = \frac{1}{2} m\omega_r^2 r^2$$

This means that when solving for the harmonic frequencies (or trap frequencies when considering optical traps for atoms), the frequencies are given as:

$$\omega_r = \sqrt{\frac{8P_0}{\pi m w_0^4}}$$

$$\omega_z = \sqrt{\frac{8P_0 \lambda^2}{m\pi^3 w_0^6}}$$

So that the relative trap frequencies for the radial and axial directions as a function of only beam waist scale as:

$$\frac{\omega_r}{\omega_z} = \frac{\pi w_0}{\lambda}$$

Setups

The most basic optical tweezer setup will likely include the following components: a laser (usually nd:yag), a beam expander, some optics used to steer the beam location in the sample plane, a microscope objective and condenser to create the trap in the sample plane, a position detector (e.G. Quadrant photodiode) to measure beam displacements and a microscope illumination source coupled to a ccd camera.

An nd:yag laser (1064 nm wavelength) is a common choice of laser for working with biological specimens. This is because such specimens (being mostly water) have a low absorption coefficient at this wavelength. A low absorption is advisable so as to minimise damage to the biological material, sometimes referred to as opticution. Perhaps the most important consideration in optical tweezer design is the choice of the objective. A stable trap requires that the gradient force, which is dependent upon the numerical aperture (na) of the objective, be greater than the scattering force. Suitable objectives typically have an na between 1.2 And 1.4.

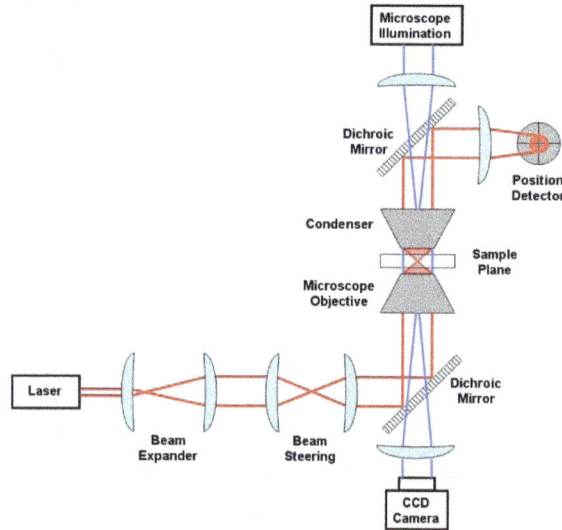

A generic optical tweezer diagram with only the most basic components.

While alternatives are available, perhaps the simplest method for position detection involves imaging the trapping laser exiting the sample chamber onto a quadrant photodiode. Lateral deflections of the beam are measured similarly to how it is done using atomic force microscopy (afm).

Expanding the beam emitted from the laser to fill the aperture of the objective will result in a tighter, diffraction-limited spot. While lateral translation of the trap relative to the sample can be accomplished by translation of the microscope slide, most tweezer setups have additional optics designed to translate the beam to give an extra degree of translational freedom. This can be done by translating the first of the two lenses labelled as "beam steering" in the figure. For example, translation of that lens in the lateral plane will result in a laterally deflected beam from what is drawn in the figure. If the distance between the beam steering lenses and the objective is chosen properly, this will correspond to a similar deflection before entering the objective and a resulting *lateral translation in the sample plane. The position of the beam waist, that is the focus of the optical trap, can be adjusted by an axial displacement of the initial lens. Such an axial displacement causes the beam to diverge or converge slightly, the end result of which is an axially displaced position of the beam waist in the sample chamber.*

Visualization of the sample plane is usually accomplished through illumination via a separate light source coupled into the optical path in the opposite direction using dichroic mirrors. This light is incident on a ccd camera and can be viewed on an external monitor or used for tracking the trapped particle position via video tracking.

Alternative Laser Beam Modes

The majority of optical tweezers make use of conventional temoo gaussian beams. However a number of other beam types have been used to trap particles, including high order laser beams i.E. Hermite-gaussian beams (temxy), laguerre-gaussian (lg) beams (templ) and bessel beams.

Optical tweezers based on laguerre-gaussian beams have the unique capability of trapping particles that are optically reflective and absorptive. Laguerre-gaussian beams also possess a well-defined orbital angular momentum that can rotate particles. This is accomplished without external mechanical or electrical steering of the beam.

Both zero and higher order bessel beams also possess a unique tweezing ability. They can trap and rotate multiple particles that are millimeters apart and even around obstacles.

Micromachines can be driven by these unique optical beams due to their intrinsic rotating mechanism due to the spin and orbital angular momentum of light.

Multiplexed Optical Tweezers

A typical setup uses one laser to create one or two traps. Commonly, two traps are generated by splitting the laser beam into two orthogonally polarized beams. Optical tweezing operations with more than two traps can be realized either by time-sharing a single laser beam among several optical tweezers, or by diffractively splitting the beam into multiple traps. With acousto-optic deflectors or galvanometer-driven mirrors, a single laser beam can be shared among hundreds of optical tweezers in the focal plane, or else spread into an extended one-dimensional trap. Specially designed diffractive optical elements can divide a single input beam into hundreds of continuously illuminated traps in arbitrary three-dimensional configurations. The trap-forming hologram also can specify the mode structure of each trap individually, thereby creating arrays of optical vortices, optical tweezers, and holographic line traps, for example. When implemented with a spatial light modulator, such holographic optical traps also can move objects in three dimensions.

Single Mode Optical Fibers

The standard fiber optical trap relies on the same principle as the optical trapping, but with the gaussian laser beam delivered through an optical fiber. If one end of the optical fiber is molded into a lens-like facet, the nearly gaussian beam carried by a single mode standard fiber will be focused at some distance from the fiber tip. The effective numerical aperture of such assembly is usually not enough to allow for a full 3d optical trap but only for a 2d trap (optical trapping and manipulation of objects will be possible only when, e.G., They are in contact with a surface). A true 3d optical trapping based on a single fiber, with a trapping point which is not in nearly contact with the fiber tip, has been realized based on a not-standard annular-core fiber arrangement and a total-internal-reflection geometry.

On the other hand, if the ends of the fiber are not moulded, the laser exiting the fiber will be diverging and thus a stable optical trap can only be realised by balancing the gradient and the scattering force from two opposing ends of the fiber. The gradient force will trap the particles in the transverse direction, while the axial optical force comes from the scattering force of the two counter propagating beams emerging from the two fibers. The equilibrium z-position of such a trapped bead is where the two scattering forces equal each other. This work was pioneered by a. Constable *et al.*, *Opt. Lett.* 18,1867 (1993), And followed by j.Guck *et al.*, *Phys. Rev. Lett.* 84, 5451 (2000), Who made use of this technique to stretch microparticles. By manipulating the input power into the two ends of the fiber, there will be an increase of a "optical stretching" that can be used to measure viscoelastic properties of cells, with sensitivity sufficient to distinguish between different individual cytoskeletal phenotypes. I.E. Human erythrocytes and mouse fibroblasts. A recent test has seen great success in differentiating cancerous cells from non-cancerous ones from the two opposed, non-focused laser beams.

Multimode Fiber-Based Traps

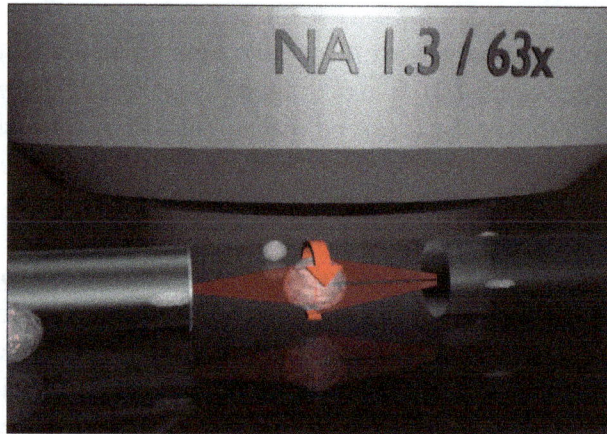

The optical cell rotator is a fiber based laser trap that can hold and precisely orient living cells for tomographic microscopy.

While earlier version of fiber-based laser traps exclusively used single mode beams, m. Kreysing and colleagues recently showed that the careful excitation of further optical modes in a short piece of optical fiber allows the realization of non-trivial trapping geometries. By this the researchers were able to orient various human cell types (individual cells and clusters) on a microscope. The main advantage of the so-called "optical cell rotator" technology over standard optical tweezers is the decoupling of trapping from imaging optics. This, its modular design, and the high compatibility of divergent laser traps with biological material indicates the great potential of this new generation of laser traps in medical research and life science. Recently, the optical cell rotator technology was implemented on the basis of adaptive optics, allowing to dynamically reconfigure the optical trap during operation and adapt it to the sample.

Cell Sorting

One of the more common cell-sorting systems makes use of flow cytometry through fluorescent imaging. In this method, a suspension of biologic cells is sorted into two or more containers, based upon specific fluorescent characteristics of each cell during an assisted flow. By using an electrical charge that the cell is "trapped" in, the cells are then sorted based on the fluorescence intensity measurements. The sorting process is undertaken by an electrostatic deflection system that diverts cells into containers based upon their charge.

In the optically actuated sorting process, the cells are flowed through into an optical landscape i.E. 2D or 3d optical lattices. Without any induced electrical charge, the cells would sort based on their intrinsic refractive index properties and can be re-configurability for dynamic sorting. An optical lattice can be created using diffractive optics and optical elements.

On the other hand, k. Ladavac *et al. Used a spatial light modulator to project an intensity pattern to enable the optical sorting process.* K. Xiao and d. G. Grier applied holographic video microscopy to demonstrate that this technique can sort colloidal spheres with part-per-thousand resolution for size and refractive index.

The main mechanism for sorting is the arrangement of the optical lattice points. As the cell flow through the optical lattice, there are forces due to the particles drag force that is competing directly with the optical gradient force *from the optical lattice point. By shifting the arrangement of the optical lattice point, there is a preferred optical path where the optical forces are dominant and biased. With the aid of the flow of the cells, there is a resultant force that is directed along that preferred optical path. Hence, there is a relationship of the flow rate with the optical gradient force. By adjusting the two forces, one will be able to obtain a good optical sorting efficiency.*

Competition of the forces in the sorting environment need fine tuning to succeed in high efficient optical sorting. The need is mainly with regards to the balance of the forces; drag force due to fluid flow and optical gradient force due to arrangement of intensity spot.

Scientists at the university of st. Andrews have received considerable funding from the uk engineering and physical sciences research council (epsrc) for an optical sorting machine. This new technology could rival the conventional fluorescence-activated cell sorting.

Evanescent Fields

An evanescent field is a residue optical field that "leaks" during total internal reflection. This "leaking" of light fades off at an exponential rate. The evanescent field has found a number of applications in nanometer resolution imaging (microscopy); optical micro-manipulation (optical tweezers) are becoming ever more relevant in research.

In optical tweezers, a continuous evanescent field can be created when light is propagating through an optical waveguide (multiple total internal reflection). The resulting evanescent field has a directional sense and will propel microparticles along its propagating path. This work was first pioneered by s. Kawata and t. Sugiura, in 1992, who showed that the field can be coupled to the particles in proximity on the order of 100 nanometers.

This direct coupling of the field is treated as a type of photon tunnelling across the gap from prism to microparticles. The result is a directional optical propelling force.

A recent updated version of the evanescent field optical tweezers makes use of extended optical landscape patterns to simultaneously guide a large number of particles into a preferred direction without using a waveguide. It is termed as lensless optical trapping ("lot"). The orderly movement of the particles is aided by the introduction of ronchi ruling that creates well-defined optical potential wells (replacing the waveguide). This means that particles are propelled by the evanescent field while being trapped by the linear bright fringes. At the moment, there are scientists working on focused evanescent fields as well.

Another approach that has been recently proposed makes use of surface plasmons, which is an enhanced evanescent wave localized at a metal/dielectric interface. The enhanced force field experienced by colloidal particles exposed to surface plasmons at a flat metal/dielectric interface has been for the first time measured using a photonic force microscope, the total force magnitude being found 40 times stronger compared to a normal evanescent wave. By patterning the surface with gold microscopic islands it is possible to have selective and parallel trapping in these islands. The forces of the latter optical tweezers lie in the femtonewton range.

The evanescent field can also be used to trap cold atoms and molecules near the surface of an optical waveguide or optical nanofiber.

Indirect Approach

Ming wu, a uc berkeley professor of electrical engineering and computer sciences invented the new optoelectronic tweezers.

Wu transformed the optical energy from low powered light emitting diodes (led) into electrical energy via a photoconductive surface. The idea is to allow the led to switch on and off the photoconductive material via its fine projection. As the optical pattern can be easily transformable through optical projection, this method allows a high flexibility of switching different optical landscapes.

The manipulation/tweezing process is done by the variations between the electric field actuated by the light pattern. The particles will be either attracted or repelled from the actuated point due to the its induced electrical dipole. Particles suspended in a liquid will be susceptible to the electrical field gradient, this is known as dielectrophoresis.

One clear advantage is that the electrical conductivity is different between different kinds of cells. Living cells have a lower conductive medium while the dead ones have minimum or no conductive medium. The system may be able to manipulate roughly 10,000 cells or particles at the same time.

Comments by professor kishan dholakia on this new technique, k. Dholakia, *nature materials 4, 579–580 (01 aug 2005) news and views:*

"The system was able to move live e. Coli bacteria and 20-micrometre-wide particles, using an optical power output of less than 10 microwatts. This is one-hundred-thousandth of the power needed for [direct] optical tweezers".

Optical Binding

When a cluster of microparticles are trapped within a monochromatic laser beam, the organization of the microparticles within the optical trapping is heavily dependent on the redistributing of the optical trapping forces amongst the microparticles. This redistribution of light forces amongst the cluster of microparticles provides a new force equilibrium on the cluster as a whole. As such we can say that the cluster of microparticles are somewhat bound together by light. One of the first evidence of optical binding was reported by michael m. Burns, jean-marc fournier, and jene a. Golovchenko .

Laser Beam Welding

A robot performs remote fibre laser welding.

Laser beam welding (lbw) is a welding technique used to join multiple pieces of metal through the use of a laser. The beam provides a concentrated heat source, allowing for narrow, deep welds and high welding rates. The process is frequently used in high volume applications using automation, such as in the automotive industry. It is based on keyhole or penetration mode welding.

Operation

Like electron beam welding (ebw), laser beam welding has high power density (on the order of 1 mw/cm2) resulting in small heat-affected zones and high heating and cooling rates. The spot size of the laser can vary between 0.2 Mm and 13 mm, though only smaller sizes are used for welding. The depth of penetration is proportional to the amount of power supplied, but is also dependent on the location of the focal point: penetration is maximized when the focal point is slightly below the surface of the workpiece

A continuous or pulsed laser beam may be used depending upon the application. Millisecond-long pulses are used to weld thin materials such as razor blades while continuous laser systems are employed for deep welds.

Lbw is a versatile process, capable of welding carbon steels, hsla steels, stainless steel, aluminum, and titanium. Due to high cooling rates, cracking is a concern when welding high-carbon steels. The weld quality is high, similar to that of electron beam welding. The speed of welding is proportional to the amount of power supplied but also depends on the type and thickness of the workpieces. The high power capability of gas lasers make them especially suitable for high volume applications. Lbw is particularly dominant in the automotive industry.

Some of the advantages of lbw in comparison to ebw are as follows:

- The laser beam can be transmitted through air rather than requiring a vacuum

- The process is easily automated with robotic machinery

- X-rays are not generated

- Lbw results in higher quality welds

A derivative of lbw, laser-hybrid welding, combines the laser of lbw with an arc welding method such as gas metal arc welding. This combination allows for greater positioning flexibility, since gmaw supplies molten metal to fill the joint, and due to the use of a laser, increases the welding speed over what is normally possible with gmaw. Weld quality tends to be higher as well, since the potential for undercutting is reduced.

Equipment

Automation and Cam

Although laser beam welding can be accomplished by hand, most systems are automated use a system of computer aided manufacturing based off of computer aided designs. Laser welding can also be coupled with milling to form a finished part.

Recently the reprap project, which historically worked on fused filament fabrication, expanded to development of open source laser welding systems. Such systems have been fully characterized and can be used in a wide scale of applications while reducing conventional manufacturing costs.

Lasers

- The two types of lasers commonly used are solid-state lasers (especially ruby lasers and nd:yag lasers) and gas lasers.

- The first type uses one of several solid media, including synthetic ruby (chromium in aluminum oxide), neodymium in glass (nd:glass), and the most common type, neodymium in yttrium aluminum garnet (nd:yag).

- Gas lasers use mixtures of gases such as helium, nitrogen, and carbon dioxide (co2 laser) as a medium.

- Regardless of type, however, when the medium is excited, it emits photons and forms the laser beam.

Solid State Laser

Solid-state lasers operate at wavelengths on the order of 1 micrometer, much shorter than gas lasers, and as a result require that operators wear special eyewear or use special screens to prevent retina damage. Nd:yag lasers can operate in both pulsed and continuous mode, but the other types are limited to pulsed mode. The original and still popular solid-state design is a single crystal shaped as a rod approximately 20 mm in diameter and 200 mm long, and the ends are ground flat. This rod is surrounded by a flash tube containing xenon or krypton. When flashed, a pulse of light lasting about two milliseconds is emitted by the laser. Disk shaped crystals are growing in popularity in the industry, and flashlamps are giving way to diodes due to their high efficiency. Typical power output for ruby lasers is 10–20 w, while the nd:yag laser outputs between 0.04–6,000 W. To deliver the laser beam to the weld area, fiber optics are usually employed.

Gas Laser

Gas lasers use high-voltage, low-current power sources to supply the energy needed to excite the gas mixture used as a lasing medium. These lasers can operate in both continuous and pulsed mode, and the wavelength of the co2 gas laser beam is 10.6 Mm, deep infrared, i.E. 'Heat'. Fiber optic cable absorbs and is destroyed by this wavelength, so a rigid lens and mirror delivery system is used. Power outputs for gas lasers can be much higher than solid-state lasers, reaching 25 kw.

Fiber Laser

In fiber lasers, the gain medium is the optical fiber itself. They are capable of power up to 50 kw and are increasingly being used for robotic industrial welding.

Laser Beam Delivery

Modern laser beam welding machines can be grouped into two types. In the traditional type, the laser output is moved to follow the seam. This is usually achieved with a robot. In many modern applications, remote laser beam welding is used. In this method, the laser beam is moved along the seam with the help of a laser scanner, so that the robotic arm does not need to follow the seam any more. The advantages of remote laser welding are the higher speed and the higher precision of the welding process.

Laser Beam Machining

Laser beam machining (lbm) is a non-traditional subtractive manufacturing process, a form of machining, in which a laser is directed towards the work piece for machining. This process uses thermal energy to remove material from metallic or nonmetallic surfaces. The laser is focused onto the surface to be worked and the thermal energy of the laser is transferred to the surface, heating and melting or vaporizing the material. Laser beam machining is best suited for brittle materials with low conductivity, but can be used on most materials.

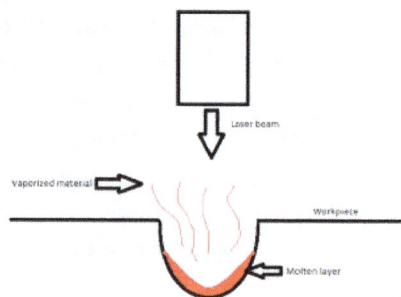

A visual of how laser beam machining works

Types of Lasers

There are many different types of lasers including gas, solid states lasers, and excimer.

In gas lasers, an electric current is liberated from a gas to generate a consistent light. Some of the most commonly used gases consist of; he-ne, ar, and co2. Fundamentally, these gases act as a pumping medium to ensure that the necessary population inversion is attained.

Solid state lasers are designed by doping a rare element into various host materials. Unlike in gas lasers, solid state lasers are pumped optically by flash lamps or arch lamps. Ruby is one of the frequently used host materials in this type of laser. A ruby laser is a type of the solid state laser whose laser medium is a synthetic ruby crystal. These ruby lasers generate deep red light pulses of a millisecond pulse length and a wavelength of about 694.3 Nm. The synthetic ruby rod is optically pumped using a xenon flashtube before it is used as an active laser medium.

Yag is an abbreviation for yttrium aluminum garnet which are crystals that are used for solid-state lasers while nd:yag refers to neodymium-doped yttrium aluminum garnet crystals that are used in the solid-state lasers as the laser medium. The nd:yag lasers emit a wavelength of light waves with high energy. Nd:glass is neodymium–doped gain media made of either silicate or phosphate materials that are used in fiber laser.

In excimer lasers, the state is different than in solid state or gas lasers. The device utilizes a combination of reactive and inert gases to produce a beam. This machine is sometimes known as an ultraviolet chemical laser.

Cutting Depth

The cutting depth of a laser is directly proportional to the quotient obtained by dividing the power of the laser beam by the product of the cutting velocity and the diameter of the laser beam spot.

$$t \propto \frac{P}{vd},$$

Where t is the depth of cut, p is the laser beam power, v is the cutting velocity, and d is the laser beam spot diameter.

The depth of the cut is also influenced by the workpiece material. The material's reflectivity, density, specific heat, and melting point temperature all contribute to the lasers ability to cut the workpiece.

The following table shows the ability of different lasers to cut different materials:

material	laser: 10.6	wavelength (micrometer) Nd:YAG laser: 1.06
ceramics	well	poorly
plywood	very well	fairly well
polycarbonate	well	fairly well
polyethylene	very well	fairly well
Perspex	very well	fairly well

Titanium	well	well
Gold	not possible	well
Copper	poorly	well
Aluminium	well	well
stainless steel		very well
construction steel		very well

Applications

Lasers can be used for welding, cladding, marking, surface treatment, drilling, and cutting among other manufacturing processes. It is used in the automobile, shipbuilding, aerospace, steel, electronics, and medical industries for precision machining of complex parts. Laser welding is advantageous in that it can weld at speeds of up to 100 mm/s as well as the ability to weld dissimilar metals. Laser cladding is used to coat cheep or weak parts with a harder material in order to improve the surface quality. Drilling and cutting with lasers is advantageous in that there is little to no wear on the cutting tool as there is no contact to cause damage. Milling with a laser is a three dimensional process that requires two lasers, but drastically cuts costs of machining parts. Lasers can be used to change the surface properties of a workpiece.

Laser beam machining can also be used in conjunction with traditional machining methods. By focusing the laser ahead of a cutting tool the material to be cut will be softened and made easier to remove, reducing cost of production and wear on the tool while increasing tool life.

The appliance of laser beam machining varies depending on the industry. In heavy manufacturing laser beam machining is used for cladding and drilling, spot and seam welding among others. In light manufacturing the machine is used to engrave and to drill other metals. In the electronic industry laser beam machining is used for wire stripping and skiving of circuits. In the medical industry it is used for cosmetic surgery and hair removal.

Advantages

Since the rays of a laser beam are monochromatic and parallel it can be focused to a very small diameter and can produce energy as high as 100 mw of energy for a square millimeter of area. It is especially suited to making accurately placed holes.

Laser beam machining has the ability to engrave or cut nearly all materials, where traditional cutting methods may fall short. There are several types of lasers, and each have different uses. For instance, materials that cannot be not be cut by a gas laser may be cut by an excimer laser.

The cost of maintaining lasers is moderately low due to the low rate of wear and tear, as there is no physical contact between the tool and the workpiece.

The machining provided by laser beams is high precision, and most of these processes do not require additional finishing.

Laser beams can be paired with gases to help the cutting process be more efficient, help minimize oxidization of surfaces, and/or keep the workpiece surface free from melted or vaporized material.

Disadvantages

The initial cost of acquiring a laser beam is moderately high. There are many accessories that aid in the machining process, and as most of these accessories are as important as the laser beam itself the startup cost of machining is raised further.

Handling and maintaining the machining requires highly trained individuals. Operating the laser beam is comparatively technical, and services from an expert may be required.

Laser beams are not designed to produce mass metal processes. For this reason production is always slow, especially when the metal processes involve a lot of cutting.

Laser beam machining consumes a lot of energy.

Deep cuts are difficult with workpieces with high melting points and usually cause a taper.

Laser Diffraction Analysis

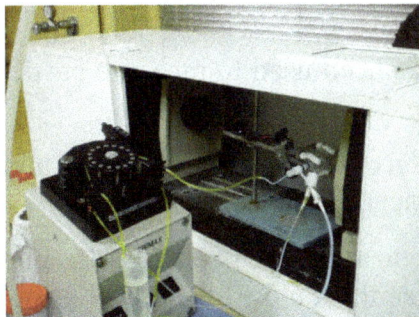

Laser diffraction analyser

Laser diffraction analysis, also known as laser diffraction spectroscopy, is a technology that utilizes diffraction patterns of a laser beam passed through any object ranging from nanometers to millimeters in size to quickly measure geometrical dimensions of a particle. This process does not depend on volumetric flow rate, the amount of particles that passes through a surface over time.

Operation

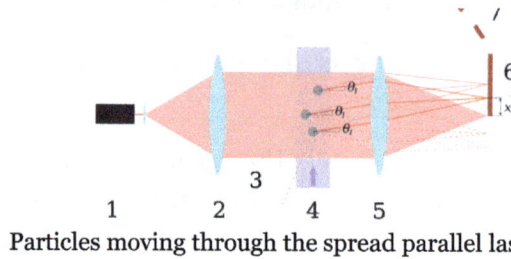

Particles moving through the spread parallel laser beam

Laser diffraction analysis is based on the fraunhofer diffraction theory, stating that the intensity of light scattered by a particle is directly proportional to the particle size. The angle of the laser beam and particle size have an inversely proportional relationship, where the laser beam angle increases as particle size decreases and vice versa.

Laser diffraction analysis is accomplished via a red he-ne laser, a commonly used gas laser for physics experiments that is made up of a laser tube, a high-voltage power supply, and structural packaging. Alternatively, blue laser diodes or leds of shorter wavelength may be used. Angling of the light energy produced by the laser is detected by having a beam of light go through a suspension and then onto a sensor. A lens is placed between the object being analyzed and the detector's focal point, causing only the surrounding laser diffraction to appear. The sizes the laser can analyze depend on the lens' focal length, the distance from the lens to its point of focus. As the focal length increases, the area the laser can detect increases as well, displaying a proportional relationship. A computer can then be used to detect the object's particle sizes from the light energy produced and its layout, which the computer derives from the data collected on the particle frequencies and wavelengths.

Uses

Laser diffraction analysis has been used to measure particle-size objects in situations such as:

- Observing distribution of sediments such as clay and mud, with an emphasis on silt and the sizes of bigger samples of clay.

- Determining in situ measurements of particles in estuaries. Particles in estuaries are important as they allow for natural or pollutant chemical species to move around with ease. The size, density, and stability of particles in estuaries are important for their transportation. Laser diffraction analysis is used here to compare particle size distributions to support this claim as well as find cycles of change in estuaries that occur because of different particles.

- Soil and its stability when wet. The stability of soil aggregation (clumps held together by moist clay) and clay dispersion (clay separating in moist soil), the two different states of soil in the cerrado savanna region, were compared with laser diffraction analysis to determine if plowing had an effect on the two. Measurements were made before plowing and after plowing for different intervals of time. Clay dispersion turned out to not be affected by plowing while soil aggregation did.

Comparisons

Since laser diffraction analysis is not the sole way of measuring particles it has been compared to the sieve-pipette method, which is a traditional technique for grain size analysis. When compared, results showed that laser diffraction analysis made fast calculations that were easy to recreate after a one-time analysis, did not need large sample sizes, and produced large amounts of data. Results can easily be manipulated because the data is on a digital surface. Both the sieve-pipette method and laser diffraction analysis are able to analyze minuscule objects, but laser diffraction analysis resulted in having better precision than its counterpart method of particle measurement.

Criticism

Laser diffraction analysis has been questioned in validity in the following areas:

- Assumptions including particles having random configurations and volume values. In some dispersion units, particles have been shown to align themselves together rather than have a turbulent flow, causing them to lead themselves in an orderly direction.

- Algorithms used in laser diffraction analysis are not thoroughly validated. Different algorithms are used at times to have collected data match assumptions made by users as an attempt to avoid data that looks incorrect.

- Measurement inaccuracies due to sharp edges on objects. Laser diffraction analysis has the chance of detecting imaginary particles at sharp edges because of the large angles the lasers make upon them.

- When compared to the data collecting of optical imaging, another particle-sizing technique, correlation between the two was poor for non-spherical particles. This is due to the fact that the underlying fraunhofer and mie theories only cover spherical particles. Non-spherical particles cause more diffuse scatter patterns and are more difficult to interpret. Some manufacturers have included algorithms in their software, which can partly compensate for non-spherical particles.

Chirped Pulse Amplification

Chirped pulse amplification (cpa) is a technique for amplifying an ultrashort laser pulse up to the petawatt level with the laser pulse being stretched out temporally and spectrally prior to amplification. Cpa is the current state of the art technique which all of the highest power lasers (greater than about 100 terawatts, with the exception of the ~500 tw national ignition facility) in the world currently utilize. Some examples of these lasers are the vulcan petawatt upgrade at the rutherford appleton laboratory's central laser facility, the diocles laser at the university of nebraska-lincoln, the gekko petawatt laser at the gekko xii facility in the institute of laser engineering at osaka university, the omega ep laser at the university of rochester's lab for laser energetics and the now dismantled petawatt line on the former nova laser at the lawrence livermore national laboratory. Apart from these state-of-the-art research systems, a number of commercial manufacturers sell ti:sapphire-based cpas with peak powers of 10 to 100 gigawatts.

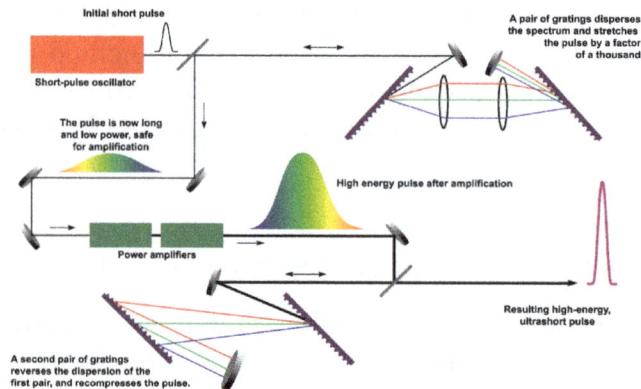

Diagramatic scheme of chirped pulse amplification.

Chirped-pulse amplification was originally introduced as a technique to increase the available power in radar in 1960. Cpa for lasers was invented by gérard mourou and donna strickland at the university of rochester in the mid 1980s. Before then, the peak power of laser pulses was limited because a laser pulse at intensities of gigawatts per square centimeter causes serious damage to the gain medium through nonlinear processes such as self-focusing. For example, some of the most powerful compressed cpa laser beams, even in an unfocused large aperture (after exiting the compression grating) can exceed intensities of 700 gigawatts/cm2, which if allowed to propagate in air or the laser gain medium would instantly self focus and form a plasma or cause filament propagation, both of which would ruin the original beam's desirable qualities and could even cause back-reflection potentially damaging the laser's components. In order to keep the intensity of laser pulses below the threshold of the nonlinear effects, the laser systems had to be large and expensive, and the peak power of laser pulses was limited to the high gigawatt level or terawatt level for very large multi beam facilities.

In cpa, on the other hand, an ultrashort laser pulse is stretched out in time prior to introducing it to the gain medium using a pair of gratings that are arranged so that the low-frequency component of the laser pulse travels a shorter path than the high-frequency component does. After going through the grating pair, the laser pulse becomes positively chirped, that is, the high-frequency component lags behind the low-frequency component, and has longer pulse duration than the original by a factor of 103 to 105. Then the stretched pulse, whose intensity is sufficiently low compared with the intensity limit of gigawatts per square centimeter, is safely introduced to the gain medium and amplified by a factor 106 or more. Finally, the amplified laser pulse is recompressed back to the original pulse width through the reversal process of stretching, achieving orders of magnitude higher peak power than laser systems could generate before the invention of cpa.

In addition to the higher peak power, cpa makes it possible to miniaturize laser systems (the compressor being the biggest part). A compact high-power laser, known as a table-top terawatt laser (t3 laser), can be created based on the cpa technique.

Stretcher and Compressor Design

There are several ways to construct compressors and stretchers. However, a typical ti:sapphire-based chirped-pulse amplifier requires that the pulses are stretched to several hundred picoseconds, which means that the different wavelength components must experience about 10 cm difference in path length. The most practical way to achieve this is with grating-based stretchers and compressors. Stretchers and compressors are characterized by their dispersion. With *negative dispersion, light with higher frequencies (shorter wavelengths) takes less time to travel through the device than light with lower frequencies (longer wavelengths). With positive dispersion, it is the other way around. In a cpa, the dispersions of the stretcher and compressor should cancel out. Because of practical considerations, the stretcher is usually designed with positive dispersion and the compressor with negative dispersion.*

In principle, the dispersion of an optical device is a function $\tau(\omega)$, where τ is the time delay experienced by a frequency component ω. (Sometimes the phase $\phi(\omega) = 2\pi\tau(\omega)c / \lambda(\omega)$ is used, where c is the speed of light and λ is the wavelength.) Each component in the whole chain from the seed laser to the output of the compressor contributes to the dispersion. It turns out to be hard to tune the dispersions of the stretcher and compressor such that the resulting pulses are shorter than about 100 femtoseconds. For this, additional dispersive elements may be needed.

With Gratings

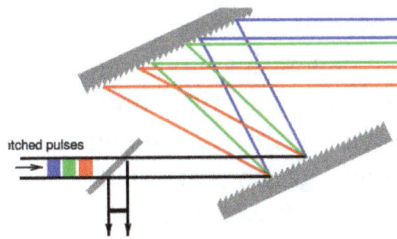

Figure 1. Schematic layout of a grating-based compressor with negative dispersion, i.E. The short wavelengths (in blue) come out first.

Figure 1 shows the simplest grating configuration, where long-wavelength components travel a larger distance than the short-wavelength components (negative dispersion). Often, only a single grating is used, with extra mirrors such that the beam hits the grating four times rather than two times as shown in the picture. This setup is normally used as a compressor, since it does not involve components that could lead to unwanted side-effects when dealing with high-intensity pulses. The dispersion can be tuned easily by changing the distance between the two gratings.

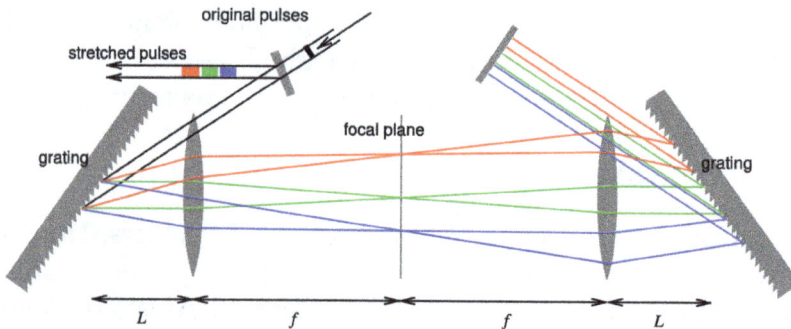

Figure 2. Schematic layout of a grating-based stretcher. In this case, $L < f$, which leads to a positive dispersion, i.E. The long wavelengths (in red) come first.

Figure 2 shows a more complicated grating configuration that involves focusing elements, here depicted as lenses. The lenses are placed at a distance $2f$ from each other (they act as a 1:1 telescope), and at a distance L from the gratings. If $L < f$, the setup acts as a *positive-dispersion stretcher and if $L > f$*, it is a negative-dispersion stretcher. And the case $L = f$ is used in the pulse shaper. Usually, the focusing element is a spherical or cylindrical mirror rather than a lens. As with the configuration in figure 1, it is possible to use an additional mirror and use a single grating rather than two separate ones. This setup requires that the beam diameter is very small compared to the length of the telescope; otherwise undesirable aberrations will be introduced. For this reason, it is normally used as a stretcher before the amplification stage, since the low-intensity seed pulses can be collimated to a beam with a small diameter.

With Prisms

It is possible to use prisms rather than gratings as a dispersive elements, as in figure 3. Despite such a simple change the set-up behaves quite differently, as to first order no group delay dispersion is introduced. Such a stretcher/compressor can have both a positive or negative dispersion, depending on the geometry and the material properties of the prisms. With lenses, the sign of the dispersion can be reversed, similar to figure 2. For a given distance between the dispersive elements, prisms generate much less dispersion than gratings. Prisms and gratings are sometimes combined to correct higher order dispersion ("grisms"), in which case the distance between the prisms is on the order of 10 meters rather than 50 cm as with a grating compressor. Gratings lose power into the other orders while prisms lose power due to rayleigh scattering.

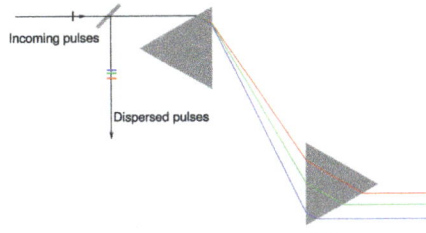

Figure 3. Prism compressor. This configuration has a positive dispersion. Although the different wavelengths appear to travel along very different paths, the effective path length differences are rather small, as indicated by the colors of the dispersed pulse.

Other Techniques

Some other techniques can be used for stretching and compressing pulses, but these are not suitable as the main stretcher/compressor in cpa due to their limited amount of dispersion and due to their inability to handle high-intensity pulses.

A pulse can be stretched simply by letting it propagate through a thick slab of transparent material, such as 200 mm glass. As with the prisms, only a limited amount of dispersion can be achieved within physically practical dimensions. Outside the visible-light spectrum, materials exist both for positive and negative dispersion. For visible and near-infrared wavelengths, almost all transparent materials have positive dispersion. However, glass fibres can have their dispersion tailored to meet the needs.

One or multiple reflections between a pair of chirped mirrors or similar device allow any form of chirp. This is often used in conjunction with the other techniques to correct for higher orders.

The dazzler is a commercial pulse shaper in which light is diffracted from an acoustic wave. By tuning the timing, frequency, and amplitude of the acoustic wave, it is possible to introduce arbitrary dispersion functions with a maximum delay of a few picoseconds.

A phase-shifting mask can be placed in the focal plane of the stretcher in fig. 2, Which introduces additional dispersion. Such a mask can be an lcd array, where the phase shift can be tuned by changing the voltage on the pixels. This can generate arbitrary dispersion functions with a maximum of a few tens of picoseconds of delay. Such a setup is called a pulse shaper.

References

- Robert Eason - Pulsed Laser Deposition of Thin Films: Applications-Led Growth of Functional Materials. Wiley-Interscience, 2006, ISBN 0471447099Kalpakjian; Schmid (2008). Manufacturing Processes for Engineering Materials (5 ed.). Prentice Hall. ISBN 9780132272711.

- Paschotta, Dr. Rüdiger. "Encyclopedia of Laser Physics and Technology - neodymium-doped gain media, laser crystals, Nd:YAG, Nd:YVO4, Nd:YLF, Nd:glass". www.rp-photonics.com. Retrieved

2016-03-01.

- Bale, A.J. (February 1987). "In situ measurement of particle size in estuarine waters" (PDF). Estuarine, Coastal and Shelf Science. 24 (2): 253–263. doi:10.1016/0272-7714(87)90068-0. Retrieved 14 November 2013.

- "Cold-Atom Physics Using Optical Nanofibres". Applied quantum physics. Vienna University of Technology. Retrieved September 10, 2012.

- "Quantum Networking with Atomic Ensembles". Caltech quantum optics. California Institute of Technology. Retrieved September 10, 2012.

- Massachusetts Institute of Technology (2007, April 8). Laser-cooling Brings Large Object Near Absolute Zero. ScienceDaily. Retrieved January 14, 2011.

Permissions

Index

www.ingramcontent.com/pod-product-compliance
Lightning Source LLC
Chambersburg PA
CBHW061948190326
41458CB00009B/2817